POWER SYSTEM MONITORING AND CONTROL

POWER SYSTEM MONITORING AND CONTROL

Hassan Bevrani

University of Kurdistan, Iran
Kyushu Institute of Technology (Visiting Professor), Japan

Masayuki Watanabe

Kyushu Institute of Technology, Japan

Yasunori Mitani

Kyushu Institute of Technology, Japan

Library of Congress Cataloging-in-Publication Data:

Bevrani, Hassan.
  Power system monitoring and control / Hassan Bevrani, Masayuki Watanabe, Yasunori Mitani.
    pages cm
  Includes bibliographical references and index.
  ISBN 978-1-118-45069-7 (hardback)
1. Electric power systems–Control.   I. Watanabe, Masayuki (Electrical engineer)   II. Mitani, Yasunori.
III. Title.
  TK1007.B483 2014
  621.31'7–dc23

                                                                        2013041134

Printed in the United States of America.

10 9 8 7 6 5 4 3 2

Dedicated to our families and our students

CONTENTS

Preface   xiii

Acknowledgments   xvii

1 AN INTRODUCTION ON POWER SYSTEM MONITORING   1

     1.1   Synchronized Phasor Measurement   2

     1.2   Power System Monitoring and Control with Wide-Area Measurements   2

     1.3   ICT Architecture Used in Wide-Area Power System Monitoring and Control   4

     1.4   Summary   5

     References   5

2 OSCILLATION DYNAMICS ANALYSIS BASED ON PHASOR MEASUREMENTS   7

     2.1   Oscillation Characteristics in Power Systems   8

         2.1.1   Eigenvalue Analysis and Participation Factor   8

         2.1.2   Oscillation Characteristics in an Interconnected Power System   9

     2.2   An Overview of Oscillation Monitoring Using Phasor Measurements   12

         2.2.1   Monitoring of the Japan Power Network   12

         2.2.2   Monitoring of the Southeast Asia Power Network   14

     2.3   WAMS-Based Interarea Mode Identification   15

     2.4   Low-Frequency Oscillation Dynamics   16

         2.4.1   Electromechanical Modes Characteristics   16

         2.4.2   Oscillation Characteristics Analyses in Southeast Asia Power Network   18

     2.5   Summary   24

     References   24

3 SMALL-SIGNAL STABILITY ASSESSMENT   26

3.1   Power System Small-Signal Stability   27
3.2   Oscillation Model Identification Using Phasor Measurements   29
      3.2.1   Oscillation Model of the Electromechanical Mode   29
      3.2.2   Dominant Mode Identification with Signal Filtering   30
3.3   Small-Signal Stability Assessment of Wide-Area Power System   32
      3.3.1   Simulation Study   32
      3.3.2   Stability Assessment Based on Phasor Measurements   33
      3.3.3   Stability Assessment Based on Frequency Monitoring   38
3.4   Summary   41
References   41

4 GRAPHICAL TOOLS FOR STABILITY AND SECURITY ASSESSMENT   43

4.1   Importance of Graphical Tools in WAMS   43
4.2   Angle–Voltage Deviation Graph   45
4.3   Simulation Results   48
      4.3.1   Disturbance in Generation Side   49
      4.3.2   Disturbance in Demand Side   50
4.4   Voltage–Frequency Deviation Graph   52
      4.4.1 ΔV–ΔF Graph for Contingency Assessment   53
      4.4.2 $\Delta V - \Delta F$ Graph for Load Shedding Synthesis   56
4.5   Frequency–Angle Deviation Graph   58
4.6   Electromechanical Wave Propagation Graph   60
      4.6.1   Wave Propagation   62
      4.6.2   Angle Wave and System Configuration   64
4.7   Summary   68
References   68

**5 POWER SYSTEM CONTROL: FUNDAMENTALS AND NEW
    PERSPECTIVES**   70

5.1   Power System Stability and Control   71
5.2   Angle and Voltage Control   73
5.3   Frequency Control   75
      5.3.1   Frequency Control Dynamic   77
      5.3.2   Operating States and Power Reserves   81
5.4   Supervisory Control and Data Acquisition   83
5.5   Challenges, Opportunities, and New Perspectives   88
      5.5.1   Application of Advanced Control Methods and Technologies   88

5.5.2   Standards Updating   90
5.5.3   Impacts of Renewable Energy Options   90
5.5.4   RESs Contribution to Regulation Services   92
5.6   Summary   94
References   95

**6 WIDE-AREA MEASUREMENT-BASED POWER SYSTEM
CONTROL DESIGN   96**
6.1   Measurement-Based Controller Design   97
6.2   Controller Tuning Using a Vibration Model   98
      6.2.1   A Vibration Model Including the Effect of Damping
              Controllers   98
      6.2.2   Tuning Mechanism   101
      6.2.3   Simulation Results   102
6.3   Wide-Area Measurement-Based Controller Design   107
      6.3.1   Wide-Area Power System Identification   107
      6.3.2   Design Procedure   110
      6.3.3   Simulation Results   110
6.4   Summary   118
References   118

7 COORDINATED DYNAMIC STABILITY AND VOLTAGE REGULATION 119
7.1   Need for AVR–PSS Coordination   120
7.2   A Survey on Recent Achievements   123
7.3   A Robust Simultaneous AVR–PSS Synthesis Approach   126
      7.3.1   Control Framework   126
      7.3.2   Developed Algorithm   128
      7.3.3   Real-Time Implementation   131
      7.3.4   Experiment Results   132
7.4   A Wide-Area Measurement-Based Coordination Approach   135
      7.4.1   High Penetration of Wind Power   136
      7.4.2   Developed Algorithm   138
      7.4.3   An Application Example   141
      7.4.4   Simulation Results   141
7.5   Intelligent AVR and PSS Coordination Design   149
      7.5.1   Fuzzy Logic-Based Coordination System   149
      7.5.2   Simulation Results   151
7.6   Summary   155
References   155

8 WIDE-AREA MEASUREMENT-BASED EMERGENCY CONTROL   **158**

8.1   Conventional Load Shedding and New Challenges   159
   8.1.1   Load Shedding: Concept and Review   159
   8.1.2   Some Key Issues   161
8.2   Need for Monitoring Both Voltage and Frequency   162
8.3   Simultaneous Voltage and Frequency-Based LS   165
   8.3.1   Proposed LS Scheme   165
   8.3.2   Implementation   167
   8.3.3   Case Studies and Simulation Results   168
   8.3.4   An Approach for Optimal UFVLS   176
   8.3.5   Discussion   177
8.4   Wave Propagation-Based Emergency Control   178
   8.4.1   Proposed Control Scheme   178
   8.4.2   Simulation Results   180
8.5   Summary   183
References   183

9 MICROGRID CONTROL: CONCEPTS AND CLASSIFICATION   **186**

9.1   Microgrids   187
9.2   Microgrid Control   192
9.3   Local Controls   195
9.4   Secondary Controls   198
9.5   Global Controls   202
9.6   Central/Emergency Controls   204
9.7   Summary   206
References   207

10 MICROGRID CONTROL: SYNTHESIS EXAMPLES   **209**

10.1   Local Control Synthesis   209
   10.1.1   Robust Voltage Control Design   209
   10.1.2   Intelligent Droop-Based Voltage and Frequency Control   215
10.2   Secondary Control Synthesis   221
   10.2.1   Intelligent Frequency Control   221
   10.2.2   ANN-Based Self-Tuning Frequency Control   228
10.3   Global Control Synthesis   235
   10.3.1   Adaptive Energy Consumption Scheduling   235
   10.3.2   Power Dispatching in Interconnected MGs   240
10.4   Emergency Control Synthesis   242

       10.4.1 Developed LS Algorithm 243
       10.4.2 Case Study and Simulation 243
   10.5 Summary 246
   References 246

Appendix A   **New York/New England 16-Machine 68-Bus System Case Study** 249

Appendix B   **Nine-Bus Power System Case Study** 254

Appendix C   **Four-Order Dynamical Power System Model and Parameters of the Four-Machine Infinite-Bus System** 256

Index 261

PREFACE

Power system monitoring and control (PSMC) is an important issue in modern electric power system design and operation. It is becoming more significant today due to the increasing size, changing structure, introduction of renewable energy sources, distributed smart/microgrids, environmental constraints, and complexity of power systems.

The wide-area measurement system (WAMS) with phasor measurement units (PMUs) provides key technologies for monitoring, state estimation, system protection, and control of widely spread power systems. A direct, more precise, and accurate monitoring can be achieved by the technique of phasor measurements and global positioning system (GPS) time signal. A proper grasp of the present state with flexible wide-area control and smart operation addresses significant elements to maintain wide-area stability in the complicated grid, with the growing penetration of distributed generation and renewable energy sources.

In response to the existing challenge of integrating advanced metering, computation, communication, and control into appropriate levels of PSMC, this book provides a comprehensive coverage of PSMC understanding, analysis, and realization. The physical constraints and engineering aspects of the PSMC have been fully considered, and developed PSMC strategies are explained using recorded real data from practical WAMS via distributed PMUs and GPS receivers in Japan and Southeast Asia (Singapore, Malaysia, and Thailand). In addition to the power system monitoring, protection, and control, the application of WAMS in emergency control schemes, as well as the control of distributed microgrids, is also emphasized.

This book will be useful for engineers and operators in power system planning and operation, as well as for academic researchers. It describes both monitoring and control issues in power systems, from introductory to advanced steps. This book can also be useful as a supplementary text for university students in electrical engineering at both undergraduate and postgraduate levels in standard courses of Power System Dynamics, Power System Analysis, and Power System Stability and Control. This book is organized into 10 chapters.

Chapter 1 introduces power system monitoring and control, especially with wide-area phasor measurement applying PMUs. Some applications of WAMS globally, as well as information and communication technology (ICT) architecture used in the phasor measurement system are outlined as an introduction.

Chapter 2 describes the oscillatory dynamics in the wide-area power system by using acquired monitoring data with phasor measurement units. Particularly, interarea low-frequency oscillations in Japan and Southeast Asia power systems have been investigated by adopting band-pass filtering based on the fast Fourier transform (FFT) technique. Since both systems have the longitudinal configuration, the low-frequency mode oscillates in the opposite phase between both ends of the power network. The oscillatory dynamics can be captured successfully by wide-area phasor measurements.

Chapter 3 emphasizes the small-signal stability assessment with phasor measurements. Particularly, the stability of the interarea low-frequency oscillation mode has been investigated by adopting the method to identify the oscillation dynamics with a simple oscillation model. The filtering approach improves the accuracy of the estimated eigenvalues. The stability can be evaluated successfully by the presented approach.

Chapter 4 introduces graphical tools for power system stability and security assessment, such as angle–voltage deviation, voltage–frequency deviation, frequency–angle deviation, and electromechanical wave propagation graphs. The necessity of using the graphical tools rather than pure analytical and mathematical approaches in wide-area power system stability and security issues is explained. Applications for designing of wide-area controllers/coordinators as well as emergency control plans are discussed.

Chapter 5 introduces the general aspects of power system stability and control. Fundamental concepts and definitions of angle, voltage, frequency stability, and existing controls are emphasized. The timescales and characteristics of various power system controls are described. The supervisory control and data acquisition (SCADA) and energy management system (EMS) architectures in modern power grids are explained. Finally, various challenges and new research directions are presented.

Chapter 6 describes a method for tuning power system stabilizers (PSSs) based on wide-area phasor measurements. The low-order system model, which holds the characteristics of the interarea oscillation mode and control unit, is identified by monitoring data from wide area phasor measurements. The effectiveness of the proposed method has been demonstrated through the power system simulation. The results show that the appropriate controller can be designed by using the identified low-order model.

Chapter 7 addresses two control strategies to achieve stability and voltage regulation, simultaneously. The first control strategy is developed using the H_∞ static output feedback control technique via an iterative linear matrix inequalities algorithm. The proposed method was applied to a four-machine infinite bus power system through a laboratory real-time experiment, and the results are compared with a conventional automatic voltage regulator–PSS design. The second control strategy uses a criterion in the normalized phase difference versus voltage deviation plane. Based on the introduced criterion, an adaptive angle-based switching strategy and negative feedback are combined to obtain a robust control methodology against load/generation disturbances.

Chapter 8 emphasizes the necessity of using both voltage and frequency data, specifically in the presence of high wind power penetration, to develop an effective load shedding scheme. First, it is shown that the voltage and frequency responses may behave in opposite directions, following contingencies, and concerning this issue, a new load shedding scheme is proposed. Then, an overview on the electromechanical waves in power systems is presented, and the amplification of a propagated wave due to reflections or in combination with waves initiated from other disturbances is studied. Finally, based on a given descriptive study of electrical measurements and electromechanical wave propagation in large electric power systems, an emergency control scheme is introduced to detect the possible plans.

Chapter 9 proposes a comprehensive review on various microgrid control loops, and relevant standards are given with a discussion on the challenges of microgrid controls. In addition to the main MG concepts, the required control loops in the microgrids are classified into primary control, secondary control, global control, and central/emergency control levels.

Chapter 10 addresses several synthesis methodology examples based on robust, intelligent, and optimal/adaptive control strategies for controller design in the microgrids. These examples cover all control levels; that is, primary, secondary, global, and central/emergency controls.

ACKNOWLEDGMENTS

Most of the contributions, outcomes, and insight presented in this book were achieved through long-term teaching and research conducted by the authors and their research groups on power system monitoring and control issues over the years. Since 2000, the authors have intensively worked on various projects in the area of power system monitoring, dynamic stability analysis, and advanced control issues.

Some previous research and project topics are power system stabilization using superconducting magnetic energy storage (Osaka University, Japan); Campus WAMS project (for the sake of monitoring of power system dynamics, stability, and power flow status using installed PMUs in several university campuses in Japan, Thailand, Malaysia, and Singapore); high-tech green campus project (Kyushu Institute of Technology and Kyushu Electric Co., Japan); intelligent/robust automatic generation control (Frontier Technology for Electrical Energy, Japan and West Regional Electric Co., Iran); power system emergency control (Queensland University of Technology, and Australian Research Council, Australia); and sophisticated smart grid controls (Kyushu Institute of Technology, Japan; and University of Kurdistan, Iran). It is a pleasure to acknowledge the support and awards the authors received from all the mentioned sources.

The authors would also like to thank their colleagues and postgraduate students Dr. M. Fathi, H. Golpira, A. G. Tikdari, S. Shokoohi, F. Habibi, N. Hajimohammadi, and R. Khezri for their active role and continuous support. Finally, the authors offer their deepest personal gratitude to their families for their patience during the preparation of this book.

1

AN INTRODUCTION ON POWER SYSTEM MONITORING

The power system is a huge system as a result of interconnections between each service area to improve reliability and economic efficiency. The social system has been mainly developed based on electrical energy for a better life and economic growth. On the other hand, the power system is exposed to the natural environment at all times; thus, the system has some small or large disturbances, for example, by lightning, storm, and apparatus faults. Under these conditions, the system should maintain stable operation to avoid blackouts in the whole system using appropriate protection and control schemes. However, in a large-scale interconnected system, there are some difficulties in evaluating and maintaining the stability of the whole system.

Recently, a new issue in power systems has come out, which is the penetration of renewable energy sources, bringing more uncertainty that requires more severe operation. For energy security, the introduction of renewable energy sources is indispensable; therefore, to maintain system reliability and make efficient use of sustainable energy, power system monitoring should be a key technology to achieve flexible operation in the system.

On the other hand, in recent years, the development of information and communication technology (ICT) has enabled more flexibility in wide-area monitoring of power systems with fast and large data transmission. Especially, the wide-area measurement

Power System Monitoring and Control, First Edition. Hassan Bevrani, Masayuki Watanabe, and Yasunori Mitani.

system (WAMS) with phasor measurement units (PMUs) is a promising technique as one of the smart grid technologies in the trunk power grid. In this chapter, basic concepts around power system monitoring are emphasized.

1.1 SYNCHRONIZED PHASOR MEASUREMENT

To monitor the power system, many measuring instruments and apparatuses are installed. Typically, the active power, reactive power, node voltage, and frequency must be monitored at all times. So far, the supervisory control and data acquisition (SCADA) system is a widely adopted monitoring system. On the other hand, the phase angle is also known as an important quantity that should be monitored for state estimation. If the phase angle can be measured, more flexible and precise monitoring could be expected.

The most important reason to measure the voltage phasor is to determine the phase reference of the measured sinusoidal voltage at all measuring points. This can be achieved by time synchronization. However, synchronized measurement is impossible by using the independent timers at the measurement points since at least 0.1 ms accuracy is required to measure the accurate phasor. On the other hand, since the global positioning system (GPS) has been opened for private use, it becomes easy to determine the precise time at a point on the globe; thus, the synchronized phasor measurement became a feasible technique.

The concept of synchronized phasor measurement was reported in the early 1980s [1]. In the literature, the method for time synchronization by GPS to calculate the phasor with high speed from the measured voltage has not been reported. In December 1993, the GPS system officially started its operation. The concept of synchronized phasor measurement using the GPS system was contributed to the *IEEE Computer Applications in Power* by Phadke [2]. Later, some advanced applications adopting phasor measurements, state estimation, instability monitoring, adaptive relay, controller tuning, and so on, were introduced. In 1998, the IEEE Standard for Synchrophasors for Power Systems was issued [3]. The synchronized phasor measurement system has been applied to the trunk power grid since the middle of the 1990s, especially in Europe and the United States. The synchronized phasor measurement can be considered as a powerful means for monitoring wide-area power systems. Some cases of worldwide blackouts have been fully monitored by the developed measurement systems [4–6].

1.2 POWER SYSTEM MONITORING AND CONTROL WITH WIDE-AREA MEASUREMENTS

Recently, some reports have been issued by the International Council on Large Electric Systems (CIGRE) on "power system security assessment" (CIGRE WG C4.601). A technical report dealing with power system monitoring on the "wide area monitoring and control for transmission capability enhancement" was also issued in 2007 [7].

In Switzerland, PMUs are installed at four substations to monitor the operating state since the system has a heavy load to the Italy system. System stability is monitored by using PMUs installed at other countries. On the other hand, in Italy, the PMUs are installed at 30 sites since the system experienced a large blackout in September 28, 2003.

There are some PMU projects at Hydro Kebec in Canada, Western Electricity Coordinating Council in the United States, and the eastern interconnected system. Virginia Tech has a project on synchronized frequency monitoring called the Frequency Monitoring Network with original measurement units.

In China, a project was started at the initiative of Tsinghua University in 1996. In the beginning, there were some issues on the communication speed and accuracy; however, the installation of PMUs has been supported since 2002. The PMUs were installed at about 88 sites by 10 new projects between 2002 and 2005. The system with functions of a graphical user interface, database, replay capability, and so on has been developed in order to monitor wide-area power system dynamics. The installation of several hundreds of PMUs has been reported at the IEEE General Meeting in 2007 [8]. The PMU is a prospective technology for the analysis of whole power system dynamics with huge networks.

In Sweden, wide-area stability is monitored by the PMUs installed at universities/institutes. Both 400 V and 400 kV nodes are measured to investigate the similarities. The monitoring network extends across three countries in Northern Europe. The interface for cooperative wide-area monitoring by sharing webpage-based online monitoring has been developed.

In Denmark, a large amount of wind generation has been accomplished, and almost all conventional generation is of the cogeneration type. The voltage and phasor of the eastern 400 and 132 kV systems are monitored to grasp the operation status for research purposes by the cooperative work of Elkraft Power Co. and the Centre for Electric Technology at Technical University of Denmark.

In Austria, the system is interconnected with many neighboring countries. The generated power at the northeast and south areas is transmitted via 200 kV lines. A number of PMUs are installed at Wien and Ternitz to monitor the power flow and temperature of transmission lines.

In Thailand, the Electricity Generating Authority of Thailand has a power system, which extends north and south, interconnected with Laos and Malaysia; therefore, the power flow is constrained by power oscillations with poor damping. The PMUs are installed at Surat Thani and Bang Saphan to monitor the state of tie-line between the central and south areas.

In Australia, the system consists of a 30 GW network of 110 and 500 kV with a distance of 5000 km. There is an issue of oscillation stability at the interarea network between the east coast and south area. The measurement network called Power Dynamic Management has been developed by the cooperation of the National Electricity Market Management Company as an independent system operator and PowerLink as the transmission company.

In Hungary, monitoring units are installed at six sites as part of the monitoring network of the Union for the Coordination of the Transmission of Electricity interconnected system called Power Log, which can measure three-phase voltage and current.

The result of monitoring interarea oscillations is used for tuning of the installed power system stabilizers.

1.3 ICT ARCHITECTURE USED IN WIDE-AREA POWER SYSTEM MONITORING AND CONTROL

It should be a very important aspect to collect the data measured by each PMU for system monitoring, state estimation, protection, and control. The measured data could be locally saved and then collected for postanalysis, or sent to a remote location in real time for system protection or real-time control. Therefore, it should be useful to know the ICT architecture. This section briefly introduces the ICT architecture used in phasor measurement systems.

Figure 1.1 shows a typical schema for a wide-area phasor measurement system including the communication and application levels. The measured data are collected by phasor data concentrators (PDCs) via a communication network. The concentrated data could be exchanged between utilities by using the standard data format including the time stamp of the synchronized GPS time. The important function of a PDC is to receive, parse, and sort incoming data frames from the multiple PMUs.

The basic requirements for a PDC are simple; however, usually the actual implementation requires a heavy computer processing task and a wideband communication. Therefore, the number of measurement units will be limited by the hardware of a PDC unit being used to handle the data concentration.

In addition, the data transmission type and communication protocols should be considered. The standards of the phasor data transmission protocol are established in IEEE C37.118. This protocol allows data-receiving devices to start and stop data flow as well as request configuration information about the sending data. Measurement systems can self-configure by requesting scaling and signal names from sending devices. This includes a notification bit to alert downstream devices for any changes in configuration.

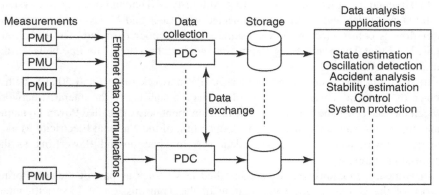

Figure 1.1. A schema of PMU/WAMS.

This protocol applies to sending data from the PMU as well as PDC devices, so it is scalable to the whole measurement system [9].

A wide variety of communication systems are used for the data collection. These include utility-owned communications, which can be a narrowband analog, digital, or wideband digital communication system. The public Internet with direct access or virtual private network technology, which is easy to implement, is now widely available. The most important aspects of choosing communications are availability, reliability, and bandwidth [9], especially for a real-time application, in which delay or interruption cannot be allowed. A narrowband communication channel could be enough to transfer data from a single measurement device. Data exchange between utilities requires a wideband communication channel since concentrated data by a PDC are accumulated data from many measurement devices.

Typical storage of data archiving for the phasor measurement system is in data files. The IEEE COMTRADE Standard (C37.110), which supports binary and floating point formats for time sequence phasor data, is widely used. The utilization of the standard file formats simplifies data exchange, analysis, and application development.

1.4 SUMMARY

This chapter introduces power system monitoring and control, especially with wide-area phasor measurement applying PMUs. Some global applications of WAMS and the ICT architecture used in the phasor measurement system have been outlined as an introduction.

REFERENCES

1. A. G. Phadke, J. S. Thorp, and M. G. Adamiak, A new measurement technique for tracking voltage phasors, local system frequency, and rate of change of frequency, *IEEE Trans. Power Apparatus Syst.*, **102**(5), 1025–1038, 1983.
2. A. G. Phadke, Synchronized phasor measurements in power systems, *IEEE Comput. Appl. Power*, **6**(2), 10–15, 1993.
3. K. E. Martin, et al., IEEE standard for synchrophasors for power systems, *IEEE Trans. Power Deliv.*, **13**(1), 73–77, 1998.
4. R. O. Burnett, Jr., M. M. Butts, T. W. Cease, V. Centeno, G. Michel, R. J. Murphy, and A. G. Phadke, Synchronized phasor measurements of a power system event, *IEEE Trans. Power Syst.*, **9**(3), 1643–1650, 1994.
5. Z. Q. Bo, G. Weller, T. Lomas, and M. A. Redfern, Positional protection of transmission systems using global positioning system, *IEEE Trans. Power Deliv.*, **15**(4), 1163–1168, 2000.
6. V. Rehtanz and D. Westermann, Wide area measurement and control system for increasing transmission capacity in deregulated energy market, In: *Proceedings of the 14th Power Systems Computation Conference*, 2002.
7. Wide area monitoring and control for transmission capability enhancement, CIGRE WG C4.601 Report, 2007.

8. Q. Yang, T. Bi, and J. Wu, WAMS implementation in China and the challenges for bulk power system protection, In: *Proceedings of the IEEE Power Engineering Society General Meeting*, 2007.

9. K. E. Martin and J. R. Carroll, Phasing in the technology, *IEEE Power Energ. Mag.*, **6**(5), 24–33, 2008.

2

OSCILLATION DYNAMICS ANALYSIS BASED ON PHASOR MEASUREMENTS

Modern interconnected wide-area power systems around the world are faced with serious challenging issues in global monitoring, stability, and control mainly due to increasing size, changing structure, emerging new uncertainties, environmental issues, and rapid growth in distributed generation. Under this circumstance, any failure in the planning, operation, protection, and control in a part of the power system could evolve into the cause of cascading events that may even lead to a large area power blackout.

These challenging issues set new demand for the development of more flexible, rapid, effective, precise, and intelligent approaches for power system dynamics monitoring, stability/security analysis, and control problems. The advent and deployment of phasor measurement units (PMUs) provides a powerful tool to enable the measurement-based methodologies for building an online dynamic snapshot-model of power systems based on real measurements to solve the mentioned problems.

This chapter introduces basic concepts of power system oscillation dynamics using phasor measurements and presents some examples for real data monitoring and analysis. Interarea low-frequency oscillations are characteristic phenomena in the interconnected power systems [1–3]. These oscillations have poor damping characteristics in heavy loading conditions on tie-lines, mainly due to the power exchange and complex power contracts under a deregulated environment. Therefore, proper estimation of the present

Power System Monitoring and Control, First Edition. Hassan Bevrani, Masayuki Watanabe, and Yasunori Mitani.

state with flexible wide-area operation and control should become key issues to keep the power system stability properly.

On the other hand, the real-time monitoring based on wide-area phasor measurements [4] attracts the attention of power system engineers for the state estimation, system protection, and control subjects [5–8]. This chapter presents a brief overview on the power system oscillation characteristics, and wide-area monitoring system (WAMS) using PMUs. To find a clear sense, the real power system in Japan and some Southeast Asian countries (Thailand, Malaysia, and Singapore) are considered as case studies. Some results for the electromechanical dynamics of real power systems are also investigated.

2.1 OSCILLATION CHARACTERISTICS IN POWER SYSTEMS

2.1.1 Eigenvalue Analysis and Participation Factor

The power swing equations of generators in an n-machine power system can be represented by [9]:

$$M_i \dot{\omega}_i = -D_i(\omega_i - 1) + P_{mi} - P_{ei}$$
$$\dot{\delta}_i = \omega_r(\omega_i - 1)$$

(2.1)

where $i = 1, 2, \ldots, n$; ω is the angular velocity; δ is the rotor angle; M is the inertia constant; D is the damping coefficient; P_m is the mechanical input to the generator; P_e is the electrical output; and ω_r is the rated angular velocity. When including the effect of other generators and controller dynamics, it is just assumed that their responses are sufficiently faster than the responses of dominant modes. Interarea oscillations are mainly caused by the swing dynamics with a large inertia represented by Equation (2.1). Now in this system suppose that a specific mode associated with power oscillation becomes unstable with variation of a parameter such as changing the loading condition. Here, consider a generator (e.g., number k) that significantly participates in the critical dominant oscillation mode. This generator can be easily selected by calculating the linear participation factor, which is defined in Reference [10].

The system dynamics is represented in general by the following equation:

$$\dot{x} = f(\mathbf{x}, p), \ \mathbf{x} \in \mathbf{R}^n, \ p \in \mathbf{R}$$

(2.2)

Linearizing (2.2) around an equilibrium point $x = x_1$ gives

$$\dot{x} = \mathbf{A}\mathbf{x}, \mathbf{A} \equiv D_x(\mathbf{x}_1, p_1)$$

(2.3)

The right eigenvector $\mathbf{u}_i$ and the left eigenvector $\mathbf{v}_i$ of the matrix $\mathbf{A}$ are defined as follows:

$$\mathbf{A}\mathbf{u}_i = \mathbf{u}_i \lambda_i$$
$$\mathbf{v}_i^T \mathbf{A} = \lambda_i \mathbf{v}_i^T$$

(2.4)

where λ_i is the ith eigenvalue of the matrix $\mathbf{A}$. It is noteworthy that the eigenvectors should be normalized to satisfy the following condition:

$$v_i^T u_j = \begin{cases} 1 & \text{if } i = j \\ 0 & \text{otherwise} \end{cases} \qquad (2.5)$$

The participation factor (p_{ki}) represents a suitable tool to measure the participation of the kth machine state in the trajectory of the ith mode. It can be defined as

$$p_{ki} = u_{ki}v_{ik} \qquad (2.6)$$

Oscillation characteristics could be explained using the participation factor [10] and the mode shape [11], which provide critical information for operational control actions. As an example, the swing characteristics of the western Japan 60 Hz system have been evaluated by calculating eigenvalues of a simulation model [2]. So far, the estimation of the participation weights has been developed based on a WAMS [12].

2.1.2 Oscillation Characteristics in an Interconnected Power System

Here, an example of the oscillation dynamics in a longitudinally interconnected power system based on the eigenvalue analysis is described. Figure 2.1 shows the West Japan 10-machine system model [13,14] that is considered in this study. The model represents a standard model for the western Japan 60 Hz power system, which was developed by the technical committee of the Institute of Electrical Engineers of Japan (IEEJ), used for the verification of simulation studies. Table 2.1 shows the system constants. Each generator is equipped with an automatic voltage regulator (AVR), which is shown in Fig. 2.2.

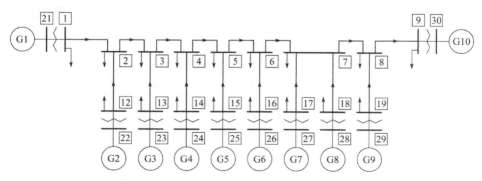

Figure 2.1. IEEJ WEST 10-machine system model.

TABLE 2.1. System Constants of WEST 10-Machine System

Generator: Park's 5th Model, 1000 MVA Base

$x_d = 1.70$ (p.u.)	$x'_d = 0.35$ (p.u.)	$x''_d = 0.25$ (p.u.)
$x_q = 1.70$ (p.u.)	$x'_q = 0.25$ (p.u.)	$M = 7.00$ (s)
$T'_d = 1.00$ (s)	$T''_d = 0.03$ (s)	$T''_q = 0.03$ (s)

Transmission System: 1000 MVA, 500 kV Base
Impedance: $Z = 0.0042 + j0.126$ (p.u.)/100 km
Electrical charge capacity: $jY/2 = j0.061$ (p.u.)/100 km
Transformer: $x_t = 0.14$ (p.u.)
Interconnected line: 100 km, double circuit
Line to generator: 50 km (G8: 100 km), double circuit

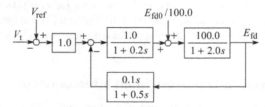

Figure 2.2. Block diagram of AVR.

In Fig. 2.2, V_t, V_{ref}, E_{fd}, and E_{fd0} are generator terminal voltage, reference voltage, AVR excitation signal, and nominal excitation signal, respectively. The rated capacity and output of the generators are shown in Table 2.2. The x_d (x_q), $x'_d(x'_q)$, and $x''_d(x''_q)$ are d-axis (q-axis) synchronous, transient, and subtransient reactance, respectively. The $T'_d(T'_q)$, and $T''_d(T''_q)$ are d-axis (q-axis) transient, and subtransient open circuit time constants, respectively.

In such a longitudinally interconnected power system, the mode associating with the low-frequency oscillation between both end generators tends to become unstable when the interconnected line is heavily loaded. Here, the load of node 2 and the power of generator 1 are increased by 1600 MW to heavily load the line between nodes 1 and 2.

TABLE 2.2. Generator Rated Capacity and Output

	Capacity, MVA	Output, MW
G1	15,000	13,500
G8	5,000	4,500
G10	30,000	27,000
Others	10,000	9,000
Total sum	120,000	108,000

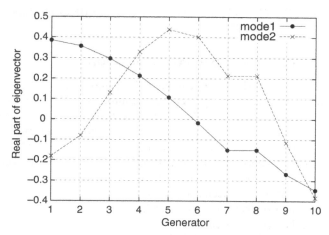

Figure 2.3. Mode shapes associated with generator angle.

This results in the destabilization of the quasi-dominant mode (mode 2) in addition to the dominant mode (mode 1).

Figures 2.3 and 2.4 show mode shapes and linear participation factors corresponding to the generator rotor angles, respectively. Figure 2.3 shows that the mode 1 oscillates in opposite phase between both end generators, while the mode 2 oscillates in opposite phase between both end group and the middle group of generators. Figure 2.4 shows that generators 1, 5, and 10 principally participate in modes 1 and 2. Therefore, it should be better to monitor both ends and the middle region of the power system to capture the characteristics of the dominant and the quasi-dominant modes.

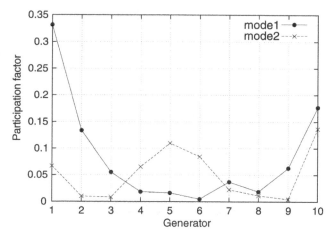

Figure 2.4. Participation factors associated with generator angle.

2.2   AN OVERVIEW OF OSCILLATION MONITORING USING PHASOR MEASUREMENTS

2.2.1   Monitoring of the Japan Power Network

In Japan, the West Japan 60 Hz system can be divided into several groups of six major electric power companies. Each group is connected through a 500 kV transmission line over a wide area. Due to its longitudinal structure there are some significant low-frequency oscillation modes in the whole system. Independent load frequency control based on tie-line bias control is adopted in each generation company [15]. Also, some independent power producers (IPPs) or power producer and suppliers are participating in the power market. So far, some oscillatory characteristics have been measured in the local area or between the interconnected areas. Here, a joint research project among some universities in Japan to develop an online wide-area measurement of power system dynamics by using the synchronized phasor measurement technique [16] is presented.

To establish a real WAMS, several PMUs are installed in universities/institutes in different geographical locations of Japan. Figure 2.5 shows the installed location of campus PMUs. The type of PMUs was NCT2000, which was manufactured by Toshiba Corp. (Fig. 2.6) [16], and synchronized by the global positioning system (GPS) signal.

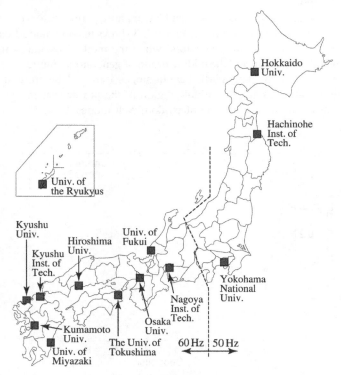

Figure 2.5. Configuration of Campus WAMS in Japan (November 2011).

Figure 2.6. Phasor measurement unit (Toshiba NCT2000).

The installation of the PMUs started in 2001 to develop a WAMS covering the whole power system in Japan as a collaborative research called Campus WAMS. At least one PMU has been installed within the service area of each power company. The PMUs measured voltage phasors of 100 V outlets, which is the standard voltage of the Japan wall outlets, in the monitoring location (laboratory) of each university campus over 24 h schedules. In practical applications, the PMUs are usually installed at substations of transmission lines.

In the developed system, the measurement interval is 2/60 s in the western 60 Hz area, and 2/50 s in the eastern 50 Hz area in order to observe the dynamic characteristics of the power swing including the local and interarea modes. Synchronized monitoring can be achieved by using the precise pulse per second (PPS) output of the GPS receiver even in the widely spread power system, and the measured remote data can be easily concentrated via a fast communication network.

Figure 2.7 shows the overall framework of the performed Campus WAMS. The measured PMU data are automatically collected by phasor data concentrators (PDCs) installed at Nagoya Institute of Technology and Kyushu Institute of Technology via the Science Information Network. Although the IEEE COMTRADE format is employed as the format of data stored in each PMU, collected data by PDCs are converted into the comma-separated value format for usability, and then converted data are stored in a network-attached storage with a large capacity.

Phasor voltage is computed using sinusoidal voltage measured at the wall outlets as follows:

$$\dot{V} = \frac{\sqrt{2}}{96} \left(\sum_{k=1}^{96} V_k \sin k\theta + j \sum_{k=1}^{96} V_k \cos k\theta \right) \tag{2.7}$$

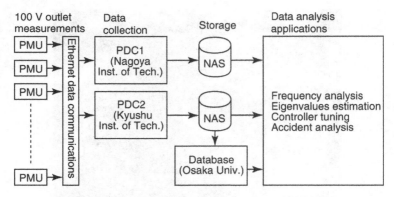

Figure 2.7. Overall schema for the Campus WAMS.

From Equation (2.7), voltage amplitude and phase can be easily obtained

$$|V| = \sqrt{\left(V_{Re}^2 + V_{Im}^2\right)} \tag{2.8}$$

$$\angle V = \tan^{-1}(V_{Im}/V_{Re}) \tag{2.9}$$

Calculating Equation (2.9) gives phase angle referred to the GPS time.

2.2.2 Monitoring of the Southeast Asia Power Network

The Thailand power system has a longitudinal configuration with an interconnection between central and southern areas by 115 and 230 kV tie-lines with an 800 km-long distance. This configuration causes the interarea low-frequency oscillations with poor damping characteristics. In Malaysia, the 500, 275, 132, and 66 kV transmission network of Tenaga Nasional Berhad (TNB) spans the whole of peninsular Malaysia, which is known as the national grid. The national grid links the electricity power producers, made up of TNB power stations and IPPs, to the TNB distribution network and some large power customers. The transmission system of Malaysia is interconnected with the Thailand system operated by the Electricity Generating Authority of Thailand in the north via a high-voltage direct current interconnection with a transmission capacity of 300 MW and a 132 kV high-voltage AC overhead line with a maximum transmission capacity of 90 MW. In the south, the Malaysia system is interconnected to the transmission system of Singapore Power at Senoko via two 230 kV submarine cables with a firm transmission capacity of 200 MW. In Singapore, the 400, 230, and 66 kV transmission network is operated by Singapore PowerGrid.

Figure 2.8 shows the developed WAMS for the power system dynamics in the Thailand, Malaysia, and Singapore power network by using the PMUs [17] (NCT2000 Type-A manufactured by Toshiba Corp. [18]). Each PMU is installed at the northern (Chiang Mai), central (Bangkok), and southern (Songkla) areas in Thailand, Kuala

Figure 2.8. The location of the installed PMUs in the Thailand, Malaysia, and Singapore power network (October 2007).

Lumpur, Malaysia, and Singapore. Measurement units are installed at domestic 220 V, 50 Hz outlets in the university campuses or company buildings. Measured phasors with time stamps synchronized with the GPS signal are collected via the Internet.

2.3 WAMS-BASED INTERAREA MODE IDENTIFICATION

In this section, a method to identify the dominant mode by using measured phasor fluctuations via the WAMS in the normal operating condition is presented. Oscillation data obtained by wide-area phasor measurements include many frequency components associated with interarea low-frequency oscillations as well as local oscillations and numerous noises. Here, dominant low-frequency oscillations are extracted by the filter to investigate the dynamics of the specified mode. A band-pass filter based on the Fourier analysis with a sharp band-pass characteristic to keep the amplitude and the phase characteristics of the original data is considered.

Discrete Fourier transform and inverse transform for finite number N of time series data x can be given by the following terms:

$$X[m] = \frac{1}{N} \sum_{n=0}^{N-1} x[n] W^{mn} \tag{2.10}$$

$$x[n] = \frac{1}{N} \sum_{m=0}^{N-1} X[m] W^{-nm} \tag{2.11}$$

where $W = \exp(-j2\pi/N)$ and m, $n = 0, 1, \ldots, N - 1$. The procedure for filtering is to hold the Fourier transform $X[m]$ of time series data $x[n]$ corresponding to the frequencies of dominant modes and eliminate $X[m]$ corresponding to the frequencies of other modes. Then, time series data of dominant modes are reconstructed by the inverse transformation (2.11). Note that this filter keeps the amplitude and the phase of extracted oscillations. The following steps summarize the procedure of the identification of the wide-area mode using a fast Fourier transform (FFT)-based filter, which provides a flexible determination method for the pass band.

Step 1: Analyze the Fourier spectrum of the phase differences,

Step 2: Determine the center frequency f_c of the band-pass filter by the spectrum in *Step 1*, and

Step 3: Extract oscillation components from original phase difference data using the FFT-based band-pass filter with $f_c \pm 0.1$ Hz.

Figure 2.9 shows an example of the filtering procedure just described. The upper part shows the original phase differences between Bangkok and Songkla (Thailand), which is a part of data measured between 15:20 and 15:40 (JST) on September 25, 2007. The center shows the filtered waveform with frequencies between 0.4 and 0.8 Hz, which includes some modes other than the dominant mode. In this case, the representative center frequency is $f_c = 0.507$ Hz, which is determined by the Fourier spectrum. The lower part shows the waveform extracted by the filter with pass band of $f_c \pm 0.1$ Hz, which oscillates with a single mode. The developed filter extracts the dominant mode, successfully.

2.4 LOW-FREQUENCY OSCILLATION DYNAMICS

Dynamic characteristics of power system oscillations can be investigated based on the measured PMU data. Numerous electromechanical modes exist in power system oscillations due to the nonlinear nature of the system. Interarea low-frequency oscillations with poor damping characteristics is a well-known problem in interconnected power systems. The characteristics of such dominant modes should be analyzed to maintain power system stability and reliability. Since, a power system shows a relatively linear behavior in the steady-state operating condition, small-signal dynamics could be investigated by using linear system concepts [3].

2.4.1 Electromechanical Modes Characteristics

Figure 2.10 shows phase differences between the monitoring stations located at the University of Miyazaki (Miyazaki, Japan) and Nagoya Institute of Technology (Nagoya, Japan), which are located at both ends of the 60 Hz Japan power system. Small-signal fluctuations caused by continuous small disturbances such as load variations are clearly observed in the measured PMU data. Many oscillation modes are superimposed;

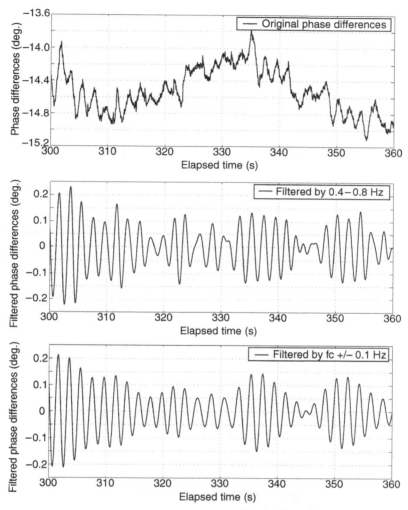

Figure 2.9. Original and filtered phase differences in Thailand.

however, a few dominant modes should be important to investigate the power system dynamics. Figure 2.11 shows the results of spectrum analysis. Low-frequency oscillations with frequency of about 0.37 Hz are also detected distinctly by phase differences between Miyazaki and Nagoya. In addition, this mode can be detected by phase differences between Hiroshima and Osaka (Japan), in the middle part of the system, although the amplitude of the mode is comparatively small. On the other hand, another quasi-dominant mode can be detected from recorded data between Miyazaki and Tokushima, where are located in one end and the central region of the system, respectively.

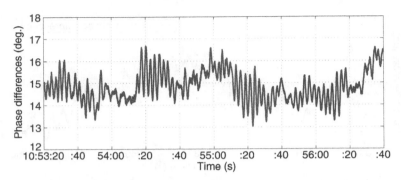

Figure 2.10. Phase difference between Miyazaki and Nagoya (August 11, 2010).

Figure 2.12 shows waveforms of most and quasi-dominant modes extracted by FFT-based filtering. These waveforms are depicted by the recorded data from Tokushima (Japan), which is roughly located in the central region of the system, as a reference of the phase angle. Figure 2.12a shows the most dominant mode (mode 1) with the frequency of about 0.37 Hz. This mode oscillates over the whole system in opposite phase with respect to another node located around the central region of the system. On the other hand, Fig. 2.12b shows waveforms of the quasi-dominant mode (mode 2) with the frequency of about 0.56 Hz, which oscillates in the same phase. It means that this swing has two nodes, where both ends oscillate in opposite phase with respect to the central region of the system. Consequently, the quasi-dominant mode has been detected in Fig 2.11c, while it has not been detected in Fig. 2.11a and b, because the mode has disappeared by recording the difference of each phase. As already described, these characteristics could be explained by the participation factor and the mode shape.

2.4.2 Oscillation Characteristics Analyses in Southeast Asia Power Network

Figure 2.13 shows frequency deviations of each area for 20 min from 15:20 (JST) on September 25, 2007. The figure shows frequency deviations of the synchronized Malaysia and Singapore system, where the Singapore system is evidently interconnected with the Malaysia system by AC transmission lines. On the other hand, the AC interconnection between the Thailand and Malaysia systems cannot be observed.

Figure 2.14 shows the frequency characteristics for frequency deviations of each area following FFT analysis. As can be seen, oscillation with a frequency of about 0.5 Hz is dominant in the Thailand system, while in the Malaysia and Singapore system, 0.3 Hz is the dominant oscillation frequency.

It is noteworthy that the spectrum of the Singapore system is larger than the other two systems. This result implies that the Singapore system mainly participates in the low-frequency mode with relatively poor damping characteristics.

Figure 2.15 shows waveforms of low-frequency oscillations extracted by the FFT-based filter. Figure 2.15a shows that the frequency deviation of the low-frequency

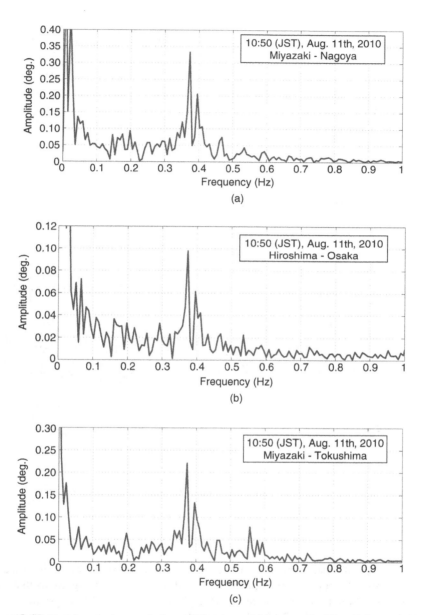

Figure 2.11. Fourier spectrum of phase differences: (a) both ends, (b) middle, and (c) end and middle.

oscillation mode in the central area (Bangkok) is oscillating in the opposite phase of the southern area (Songkla) low-frequency oscillation mode. Northern and central areas of Thailand form a coherent group with a large inertia, while the southern area oscillates against the coherent area.

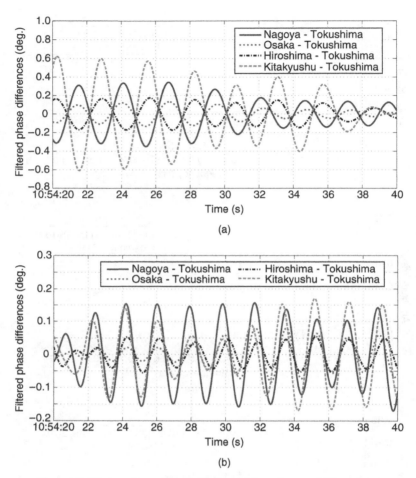

Figure 2.12. Low-frequency oscillations: (a) mode 1 and (b) mode 2.

The impact of system inertia on the frequency response is well discussed in Reference [19]. The configuration of the Thailand network is analogous to that of a single machine (Songkla) and infinite bus (Bangkok and Chiang Mai) system.

On the other hand, Fig. 2.15b shows that frequency deviations of Malaysia and Singapore oscillate in the opposite phase with each other, that is, each area seems to make a coherent group. Interarea low-frequency oscillations between each group with a frequency of about 0.3 Hz can be clearly observed. Major power plants concentrate on large cities, particularly in Kuala Lumpur and Singapore, and since these areas are interconnected with weak tie-lines, the interarea low-frequency oscillations tend to put the system in an unstable condition.

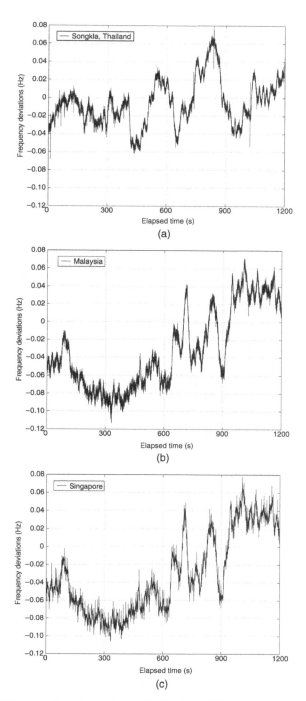

Figure 2.13. Frequency deviations during 15:20–15:40 (JST) on September 25, 2007: (a) Thailand, (b) Malaysia, and (c) Singapore.

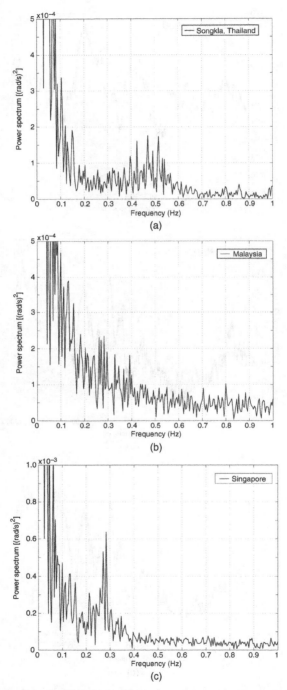

Figure 2.14. Frequency analysis: (a) Thailand, (b) Malaysia, and (c) Singapore.

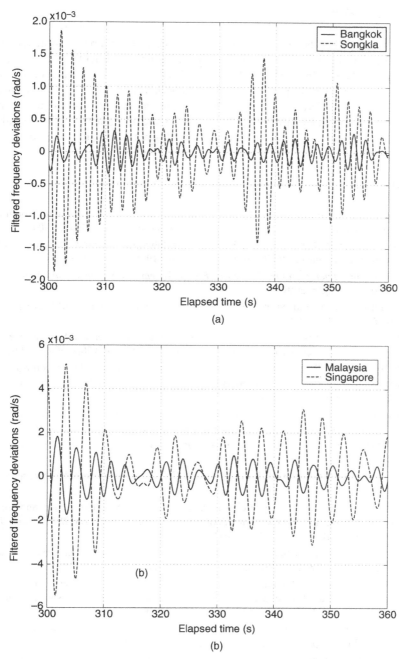

Figure 2.15. Low-frequency oscillations in Southeast Asia power systems: (a) about 0.5 Hz in Thailand and (b) about 0.3 Hz in Malaysia and Singapore.

2.5 SUMMARY

This chapter describes the oscillatory dynamics in the wide-area power system by using acquired monitoring data with phasor measurement units. Particularly, interarea low-frequency oscillations in Japan and Southeast Asia power systems have been investigated by adopting the band-pass filtering based on the fast Fourier transform technique. Since both systems have the longitudinal configuration, the low-frequency mode oscillates in the opposite phase between both ends of the power network. The oscillatory dynamics can be captured successfully by the wide-area phasor measurements.

REFERENCES

1. G. Liu, Z. Xu, Y. Huang, and W. Pan, Analysis of inter-area oscillations in the South China interconnected power system, *Elect. Power Syst. Research*, **70**(1), 38–45, 2004.

2. N. Kakimoto, A. Nakanishi, and K. Tomiyama, Instability of interarea oscillation mode by autoparametric resonance, *IEEE Trans. Power Syst.*, **19**(4), 1961–1970, 2004.

3. A. R. Messina, *Inter-area Oscillations in Power Systems: A Nonlinear and Nonstationary Perspective*, Springer, 2009.

4. A. G. Phadke, Synchronized phasor measurements in power systems, *IEEE Comput. Appl. Power*, **6**(2), 10–15, 1993.

5. J. Hauer, D. Trudnowski, G. Rogers, B. Mittelstadt, W. Litzenberger, and J. Johnson, Keeping an eye on power system dynamics, *IEEE Comput. Appl. Power*, **10**(4), 50–54, 1997.

6. C. Rehtanz and D. Westermann, Wide area measurement and control system for increasing transmission capacity in deregulated energy markets, In: *Proceedings of the 14th Power Systems Computation Conference*, 2002.

7. N. Kakimoto, M. Sugumi, T. Makino, and K. Tomiyama, Monitoring of interarea oscillation mode by synchronized phasor measurement, *IEEE Trans. Power Syst.*, **21**(1), 260–268, 2006.

8. A. G. Phadke, et al., The wide world of wide-area measurement, *IEEE Power Energ. Mag.*, **6** (5),52–65, 2008.

9. P. M. Anderson and A. A. Fouad, *Power System Control and Stability*, The Iowa State Univ. Press, Ames, IA, 1977.

10. I. J. Peez-Arriaga, G. C. Verghese, and F. C. Schweppe, Selective modal analysis with application to electric power systems, PART 1: Heuristic introduction, *IEEE Trans. Power Apparatus Syst.*, **101**(9), 3117–3125, 1982.

11. D. J. Trudnowski, Estimation electromechanical mode shape from synchrophasor measurements, *IEEE Trans. Power Syst.*, **23**(3), 1188–1195, 2008.

12. C. Li, M. Watanabe, Y. Mitani, and B. Monchusi, Participation weight estimation in power oscillation mode based on synchronized phasor measurements and auto-spectrum analysis, *IEEJ Trans. Power Energy*, **129**(12), 1449–1456, 2009.

13. Technical Committee of IEEJ, Japanese Power System Models. 1999. [Online] Available: http://www2.iee.or.jp/ver2/pes/23-st_model/english/index.html

14. M. Watanabe, Y. Mitani, and K. Tsuji, A numerical method to evaluate power system global stability determined by limit cycle, *IEEE Trans. Power Syst.*, **19**(4), 1925–1934, 2004.

15. H. Bevrani, and T. Hiyama, On load-frequency regulation with time delays: design and real-time implementation, *IEEE Trans. Energy Conversion*, **24**, 292–300, 2009.

16. T. Hashiguchi, M. Yoshimoto, Y. Mitani, O. Saeki, and K. Tsuji, Oscillation mode analysis in power systems based on data acquired by distributed phasor measurement units. In: *Proceedings of the 2003 IEEE International Symposium on Circuits and Systems*, III-367–370, 2003.

17. M. Watanabe, K. Higuma, Y. Mitani, and I. Ngamroo, Analyses of power system dynamics in Southeast Asia power network based on multiple synchronized phasor measurements. In: *Proceedings of the IEEJ-EIT Joint Symposium on Advanced Technology in Power Systems*, 2007.

18. R. Tsukui, P. Beaumont, T. Tanaka, and K. Sekiguchi, Intranet-based protection and control, *IEEE Comput. Appl. Power*, **14**(2), 14–17, 2001.

19. H. Bevrani, *Robust Power System Frequency Control*, Springer, NY, 2009.

3

SMALL-SIGNAL STABILITY ASSESSMENT

Power system stability can be affected by many factors, such as the growth of demand, the use of more large-capacity generators installed at remote locations from the demand center, longer transmission lines, the heavier power flow on tie-lines, especially the penetration of renewable energy sources with large power fluctuations, and so on. The risk of synchronous generator outage or a wide-area power blackout should be reduced even in such severe conditions. The techniques of monitoring and estimation of power system stability are key issues to prevent power outages.

In recent years, various methods of online monitoring have been proposed [1–6]. These methods place emphasis on the adaptive wide-area control to address complicated system state changes. For this purpose, in order to grasp the behavior of the whole power system in real time, wide-area data acquisition is required. To implement such real-time monitoring and control based on wide-area data acquisition, some parameters such as phase angle, bus voltage, frequency, and line power flow at multiple sites should be measured simultaneously.

This chapter introduces basic concepts of the power system small-signal stability assessment. As described in Chapter 2, interarea low-frequency oscillations can be considered as characteristic phenomena in large-scale interconnected power systems. The electromechanical dynamics of interarea oscillations with poor damping

Power System Monitoring and Control, First Edition. Hassan Bevrani, Masayuki Watanabe, and Yasunori Mitani.
© 2014 John Wiley & Sons, Inc. Published 2014 by John Wiley & Sons, Inc.

characteristics in an interconnected power system are important issues since their characteristics are dominant in power system stability, which can be monitored successfully with the PMUs.

This chapter presents an approach for the small-signal stability assessment of the interarea low-frequency oscillation based on the eigenvalues estimation for the detected dominant oscillations from the acquired phasor data.

3.1 POWER SYSTEM SMALL-SIGNAL STABILITY

Here, a general method to evaluate the stability of a dynamical system is described. The system including dynamics of all components is generally represented by

$$\frac{d}{dt}\mathbf{x} = f(\mathbf{x}) \tag{3.1}$$

where $\mathbf{x}$ is the vector of state variables. In a power system, the state vector $\mathbf{x}$ consists of the rotor angle, the speed deviation, variables associated with the electric responses of various windings of generator, variables associated with the dynamics of exciter and governor of each generator, variables associated with other control devices that have dynamic characteristics, for example, flexible alternating current transmission system (FACTS) devices, and so on. In the power system, the state equation (3.1) is highly nonlinear.

The state equation (3.1) can be linearized at the neighborhood of the specified equilibrium point:

$$\frac{d}{dt}\Delta\mathbf{x} = \frac{\partial f(\mathbf{x})}{\partial \mathbf{x}}\Delta\mathbf{x} = \mathbf{A}\Delta\mathbf{x} \tag{3.2}$$

where $\mathbf{A}$ is the Jacobian matrix

$$\mathbf{A} = \frac{\partial f(\mathbf{x})}{\partial \mathbf{x}} = \begin{bmatrix} \dfrac{\partial f_1}{\partial x_1} & \cdots & \dfrac{\partial f_1}{\partial x_n} \\ \vdots & \ddots & \vdots \\ \dfrac{\partial f_n}{\partial x_1} & \cdots & \dfrac{\partial f_n}{\partial x_n} \end{bmatrix} \tag{3.3}$$

A general solution of the state equation (3.2) can be represented as follows:

$$\mathbf{x} = c_1\mathbf{x}_1\exp(\lambda_1 t) + c_2\mathbf{x}_2\exp(\lambda_2 t) + \cdots + c_n\mathbf{x}_n\exp(\lambda_n t) \tag{3.4}$$

where $\mathbf{x}_1, \mathbf{x}_2, \ldots, \mathbf{x}_n$ are eigenvectors of the matrix $\mathbf{A}$. On the other hand, $\lambda_1, \lambda_2, \ldots, \lambda_n$ are eigenvalues of the matrix $\mathbf{A}$, which are determined to satisfy

$$[\mathbf{A} - \lambda\mathbf{I}]\mathbf{x} = 0 \tag{3.5}$$

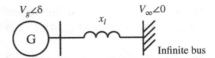

Figure 3.1. Single machine infinite bus system model.

That is, the determinant of (3.5) should be zero.

$$det[\mathbf{A} - \lambda\mathbf{I}] = 0 \qquad (3.6)$$

The condition for the stable system is that all of the real parts of eigenvalues should be negative.

As an example, here the oscillation dynamics of a single machine infinite bus system shown in Fig. 3.1 is evaluated by the eigenvalue analysis.

The swing equation of a synchronous generator is given by

$$M\frac{d}{dt}\omega + D(\omega - 1) = P_m - P_e$$
$$P_e = \frac{V_g V_\infty \sin\delta}{x'_d + x_l} \qquad (3.7)$$

where ω is the angular velocity, M is the inertia constant, D is the damping coefficient, P_m is the mechanical input to the generator, P_e is the electrical output, V_g is the generator voltage, V_∞ is the voltage of the infinite bus, δ is the rotor angle of the generator, x'_d is the transient reactance of the generator, and x_l is the line reactance. Note that the resistance of the generator and the transmission line is not considered for simplicity. In the swing equation (3.7), assuming that the voltages V_g and V_∞ are constant, and $\cos \Delta\delta \cong 1$, $\sin \Delta\delta \cong \Delta\delta$, the deviation of the output power of the generator is given by

$$\Delta P_e \cong \frac{V_g V_\infty \cos \delta_0}{x'_d + x_l}\Delta\delta \qquad (3.8)$$

where the subscript 0 is used to denote the initial value of the equilibrium point. Therefore, the swing equation (3.7) can be rewritten as

$$M\frac{d}{dt}\Delta\omega + D\Delta\omega = -\frac{V_g V_\infty \cos \delta_0}{x'_d + x_l}\Delta\delta \qquad (3.9)$$

Here, the mechanical input P_m is assumed to be constant.

Finally, the linearized swing equation can be described by the following second-order linear differential equation:

$$M\frac{d^2}{dt^2}\Delta\delta + D\frac{d}{dt}\Delta\delta + K\Delta\delta = 0$$
$$K = \frac{V_g V_\infty \cos \delta_0}{x'_d + x_l} \qquad (3.10)$$

The characteristic equation of the swing equation (3.10) is given by

$$Ms^2 + Ds + K = 0 \tag{3.11}$$

The eigenvalue is the solution of the characteristic equation (3.11)

$$s = -\frac{D}{2M} \pm j\frac{\sqrt{4MK - D^2}}{2M} \tag{3.12}$$

The time domain solution of the swing equation (3.10) is given by

$$\Delta\delta = A\exp(\alpha t)\sin(\beta t + \varphi) \tag{3.13}$$

where

$$\alpha = -\frac{D}{2M}, \beta = \frac{\sqrt{4MK - D^2}}{2M}$$

If the real part of eigenvalues α is positive, the solution diverges, that is, the generator loses the synchronism. On the other hand, if the real part of eigenvalues α is negative, the solution converges to the equilibrium point, that is, the system maintains stability. Thus, the power system stability in the vicinity of the equilibrium point can be evaluated by calculating the eigenvalues of matrix **A**. In fact, the real part of eigenvalues provides an index for the degree of stability, and the imaginary part of the eigenvalues β corresponds to the angular velocity of oscillation.

3.2 OSCILLATION MODEL IDENTIFICATION USING PHASOR MEASUREMENTS

3.2.1 Oscillation Model of the Electromechanical Mode

The power swing equations of generators in an n-machine system can be represented by [7]

$$\begin{aligned} M_i\dot{\omega}_i &= -D_i(\omega_i - 1) + P_{mi} - P_{ei} \\ \dot{\delta}_i &= \omega_r(\omega_i - 1) \end{aligned} \tag{3.14}$$

where $i = 1, 2, \ldots, n$, ω is the angular velocity, δ is the rotor angle, M is the inertia constant, D is the damping coefficient, P_m is the mechanical input to the generator, P_e is the electrical output, and ω_r is the rated angular velocity. When including the effect of other generator and controller dynamics, it is just assumed that their responses are sufficiently faster than the responses of the dominant modes. Interarea oscillations are mainly caused by the swing dynamics with a large inertia represented by (3.14). Now in this system suppose that the specific mode associated with power oscillation becomes unstable with the variation of a parameter such as changing the loading condition. The generator that significantly participates in the critical dominant oscillation mode can be easily identified by calculating the linear participation factor (2.6) [8].

As described in Chapter 2, interarea low-frequency oscillations involving both end generators are dominant in longitudinally interconnected power systems such as the western Japan 60 Hz power system. The characteristics of interarea oscillations caused by the inertia of some groups of synchronous generators could be represented by the same dynamics given by (3.14). Therefore, the interarea oscillation dynamics with a single mode can be simplified by assuming that it is analogous to a single machine and an infinite bus system:

$$
\dot{\omega} = f(\Delta\omega, \Delta\delta)
$$
$$
\dot{\delta} = \Delta\omega
$$
(3.15)

Here, the dynamics of the critical dominant mode is represented by a simplified oscillation model. A second-order oscillation model can be identified in the following form by using time series data of voltage phasors of two measurement sites:

$$
\begin{bmatrix} \dot{x}_1 \\ \dot{x}_2 \end{bmatrix} = \begin{bmatrix} a_1 & a_2 \\ 1 & 0 \end{bmatrix} \begin{bmatrix} x_1 \\ x_2 \end{bmatrix}
$$
$$
= A \begin{bmatrix} x_1 \\ x_2 \end{bmatrix}
$$
(3.16)

where $x_1 = \dot{\delta}_1 - \dot{\delta}_s$ and $x_2 = \delta_1 - \delta_s - (\delta_{1e} - \delta_{se})$; the subscript 1 denotes the selected site, the subscript s denotes the reference site, and subscript e denotes the initial phase angle. The coefficients a_1 and a_2 can be determined by the least squares method using acquired voltage phasor data. The characteristics of the dominant mode can be evaluated by the eigenvalues of the coefficient matrix $\mathbf{A}$.

Using steady-state phasor fluctuations, the model (3.16) can be identified. Hence, it is not necessary to stimulate the system by injecting a test signal such as an input step signal [9].

3.2.2 Dominant Mode Identification with Signal Filtering

Acquired voltage phasor data include many oscillation modes with different frequencies associated with interarea low-frequency oscillations with frequencies under 1 Hz, local modes with frequencies around 1 Hz, other modes with much lower frequencies associated with load frequency control issue. Therefore, the critical dominant mode should be extracted by some filtering techniques in order to estimate the simplified oscillation model (3.16) with higher accuracy. Here, two filtering methods are compared by focusing on the accuracy of the mode identification:

1. Discrete wavelet transformation (DWT)-based filtering
2. Fast Fourier transformation (FFT)-based filtering

The DWT can be used to decompose the signal into some signal components by keeping the time information. Therefore, the DWT can be an effective tool in detecting some events buried in the acquired time series data. The wavelet analysis decomposes a signal applying the shifting and scaling of the original (mother) wavelet. However, in the DWT, the passband of the signal decomposition filter is fixed since the filter is composed of a binary scale, that is, there is no flexibility to apply more detailed analysis [10].

Here, the application of the DWT to the power system oscillation data is considered. As described in Chapter 2, the interarea low-frequency oscillation mode is dominant in the whole system, and the characteristics of the mode can be identified by using the data acquired from both ends of the system since both end generators mainly participate in this mode. The oscillation mode between 0.2 and 0.8 Hz can be extracted by applying the DWT with the Symlet wavelet function.

On the other hand, the low-frequency mode might be extracted by acquired data from measurement units installed in the local area since the mode oscillates in the whole power system. It would be effective if the low-frequency mode could be detected by using only local measurements, for example, the measurements in the respective service areas since the data exchange between power companies might not be required in this case. However, the amplitude of the dominant oscillations included in the data measured in a local area should be smaller than the data measured in the wide area. That is, the influence of other modes becomes relatively larger; therefore, a specific filter with much narrower passband should be applied to extract the dominant mode and identify the dynamics with high accuracy.

Here, a band-pass filter based on the Fourier analysis with a sharp band-pass characteristic, while keeping the amplitude and the phase characteristics of the original data, is considered. Discrete Fourier transform and inverse transform for the finite number N of time series data x are given by

$$X[m] = \frac{1}{N}\sum_{n=0}^{N-1} x[n]W^{mn} \qquad (3.17)$$

$$x[n] = \frac{1}{N}\sum_{m=0}^{N-1} X[m]W^{-nm} \qquad (3.18)$$

where $W = \exp(-j2p/N)$ and $m, n = 0, 1, \ldots, N-1$. The procedure of filtering is to hold the Fourier transform $X[m]$ of time series data $x[n]$ corresponding to the frequencies of dominant modes and eliminate $X[m]$ corresponding to the frequencies of other modes, then time series data of the dominant modes are reconstructed by the inverse transformation (3.18). It is noteworthy that this filter keeps the amplitude and phase of extracted oscillations. The following steps summarize the procedure of the stability assessment of the wide-area mode using the FFT-based filtering approach:

Step 1: Analyze the Fourier spectrum of the phase differences.

Step 2: Determine the center frequency f_c of the band-pass filter by the spectrum in step 1.

Step 3: Extract oscillation components from the original phase difference data using the FFT-based band-pass filter with $f_c \pm 0.1$ Hz.

Step 4: Identify the simplified oscillation model (3.16) by using the extracted phase difference data, and then evaluate the eigenvalues of the dominant mode.

3.3 SMALL-SIGNAL STABILITY ASSESSMENT OF WIDE-AREA POWER SYSTEM

3.3.1 Simulation Study

Here, the described method for the small-signal stability assessment of the wide-area power system with phasor measurements is applied to a longitudinally interconnected six-machine power system shown in Fig. 3.2. The system constants are given in Table 2.1. Each generator is equipped with an automatic voltage regulator (AVR), which is shown in Fig. 2.2 [11]. The rated capacity of the generators is shown in Table 3.1.

Small load fluctuations measured in the steady-state operation can be used to identify (3.16), which represents the dynamics of the dominant mode. In this example, small load fluctuations are generated by slightly changing the load in order to simulate fluctuations measured in the real power system. Here, the stability assessment by the phasor measurements in the local area is considered. The simplified model (3.16) is identified by using the phase difference data between nodes 23 and 24 as adjacent measurement sites. The data length for the model identification is 200 s.

In case 1, the true eigenvalues of the dominant modes, calculated by the linearization of the system model, are $-0.09 \pm j1.93$. When the DWT-based filter with the passband from 0.2 to 0.8 Hz is applied for extracting dominant oscillations, the coefficients of the identified model are $a_1 = -0.282$ and $a_2 = -4.334$, that is, the estimated eigenvalues are

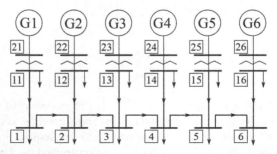

Figure 3.2. Six-machine longitudinally interconnected power system model.

TABLE 3.1. Generator Rated Capacity (MVA)

	G1	G2	G3, G5	G4	G6	Total Sum
Case 1	20,000	13,500	6,750	40,000	33,000	120,000
Case 2	16,000	10,800	5,400	32,000	27,000	96,600
Case 3	12,000	8,000	4,000	24,000	20,000	72,000

$-0.14 \pm j2.08$. The result shows that the estimated eigenvalues are a little far from the real ones. On the other hand, when the FFT-based filter with the center frequency $f_c = 0.31$ Hz and the passband between 0.19 and 0.43 Hz is applied, the coefficients of the identified model is $a_1 = -0.225$ and $a_2 = -3.730$, that is, the estimated eigenvalues are $-0.11 \pm j1.93$. More close eigenvalues can be estimated by applying the FFT-based filtering. A narrower band-pass filter can be effective in eliminating the impact of other modes, especially in extracting the dominant oscillations from phase difference between adjacent measurement sites.

Figure 3.3 shows the original and filtered oscillations. Figure 3.3a shows the original second differentiation of the phase difference between nodes 23 and 24 before filtering. The dashed lines in Fig. 3.3b and c show the extracted second differentiation of the phase difference between nodes 23 and 24 by applying the DWT-based and the FFT-based filtering approaches, respectively. The solid lines show the waveform calculated by the identified model (3.16). Figure 3.3b shows that multiple modes are included in the waveform extracted by the DWT-based filtering since the amplitude of the dominant mode becomes comparable to other modes. On the other hand, Fig. 3.3c shows that a single mode is successfully extracted by specifying the passband of the FFT-based filtering method.

Table 3.2 summarizes the comparison of eigenvalues considering the type of filters for three cases. The value in the bracket indicates the residual for the least squares. Note that each value is normalized by each maximum value since the amplitude of the extracted waveform is different in each case. All results obtained from the FFT-based filtering have smaller error than the results given by the DWT-based filtering approach. These results demonstrate that the accuracy of the estimated eigenvalues can be improved by the FFT-based filtering method.

3.3.2 Stability Assessment Based on Phasor Measurements

The method is applied to phasor data measured in the real power system. Figure 3.4 shows the location of PMUs (Toshiba NCT2000) [12] installed at the western Japan 60 Hz power system. Figure 3.5 shows the Fourier spectrum of the phase difference data between Miyazaki and Nagoya (recorded at University of Miyazaki and Nagoya Institute of Technology-N.I.T., respectively), which are located at both ends of the system, and between Hiroshima and Osaka, which are located in the middle part of the system, respectively. The data of 200 s from 04:50 on July 1, 2006 are extracted by the DWT-based filter, which has the passband between 0.2 and 0.8 Hz. When using the phase difference between both ends, oscillations around 0.4 Hz, which are the interarea low-frequency mode, can be observed dominantly. On the other hand, when using the phase

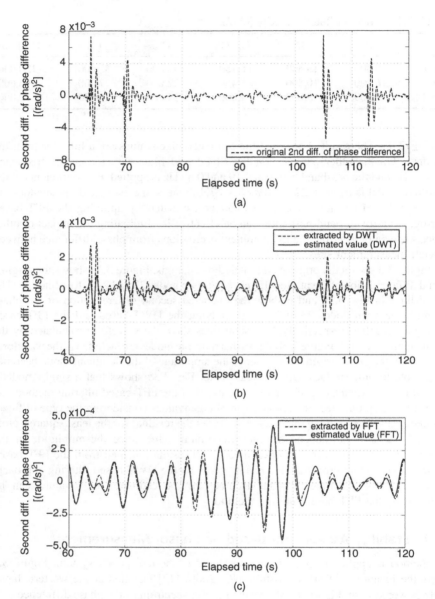

Figure 3.3. Original and filtered oscillations: (a) original, (b) DWT, and (c) FFT.

difference between adjacent sites, oscillations around 0.4 Hz can be observed; however, the amplitude becomes comparable to other modes.

Figure 3.6 shows the comparison of the accuracy of the simplified model (3.16) when the DWT- and the FFT-based filtering methods are applied to the phase difference between adjacent sites, respectively. The sum of square errors for DWT and FFT are

TABLE 3.2. Comparison of Eigenvalues with the Type of Filtering Method

	Original	DWT (Residual)	FFT (Residual)
Case 1	$-0.09 \pm j1.93$	$-0.14 \pm j2.08$ (73.1)	$-0.11 \pm j1.93$ (21.0)
Case 2	$-0.13 \pm j2.85$	$-0.17 \pm j2.78$ (46.2)	$-0.12 \pm j2.85$ (26.5)
Case 3	$-0.16 \pm j3.21$	$-0.19 \pm j3.07$ (55.1)	$-0.15 \pm j3.22$ (31.0)

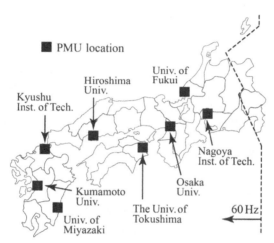

Figure 3.4. The location of installed PMUs.

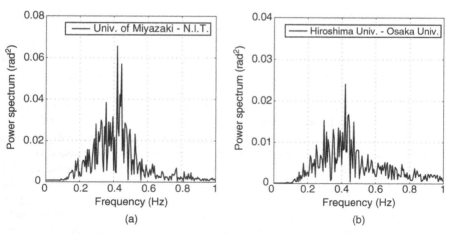

Figure 3.5. FFT results: (a) both ends and (b) middle.

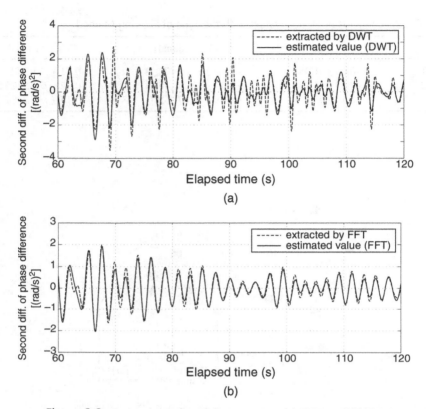

Figure 3.6. Comparison of modeling accuracy: (a) DWT and (b) FFT.

110.7 and 36.9, respectively, which are normalized by the maximum value. The accuracy of the identified model has been improved by the FFT-based filter.

Figure 3.7 shows the stability assessment of every hour for a week from July 1 to July 7 in 2006. Each eigenvalue is depicted by using 200 s data from 50 min of every hour. The reference of eigenvalues is the value estimated by using the phase difference between both ends.

Figure 3.7 shows the comparison of the results of the two mentioned filters by using phase differences between adjacent sites. In the case of the DWT-based filtering, the estimated eigenvalues have larger errors with respect to the reference eigenvalues since the accuracy of the identified model deteriorates because the phase difference data includes multiple oscillation modes. On the other hand, in the case of FFT filtering, the daily change and the difference between weekdays and holidays of the stability of the low-frequency mode can be evaluated successfully since each eigenvalue has almost the same value as the reference. Thus, the wide-area stability can be evaluated with a high accuracy when using phasor data of the adjacent sites by improving the filter processing.

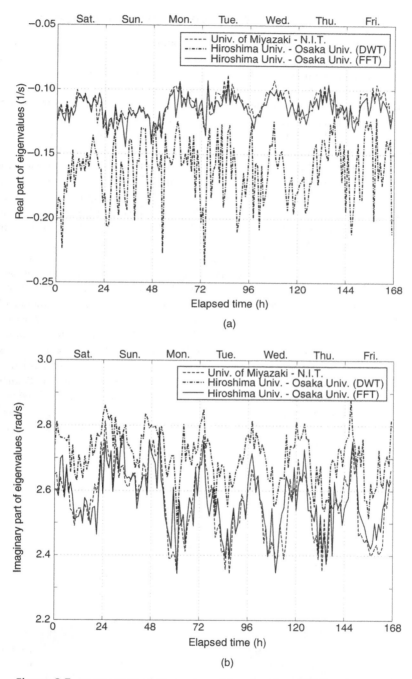

Figure 3.7. Comparison of eigenvalues: (a) real part and (b) imaginary part.

3.3.3 Stability Assessment Based on Frequency Monitoring

In a longitudinally interconnected power system, both end generators mainly participate in the low-frequency oscillation mode; therefore, the mode can be extracted from the phase difference between any two measurement sites in the system as described in the previous section. On the other hand, the mode can be extracted from the phasor information of just one measurement site located at one end since phasor oscillations of both ends have large amplitudes. Figure 3.8a shows the FFT spectrum for the phase difference between Miyazaki and Nagoya located at both ends of the system for data of 200 s from 13:50 on August 18, 2006. Figure 3.8b shows the FFT spectrum for the frequency deviations of Nagoya obtained by differentiating the measured phasors for the same period. The low-frequency oscillations with the frequency of about 0.4 Hz can be observed in both spectrums; however, Fig. 3.8b shows that the amplitude of this mode is comparable to other modes. Therefore, the DWT-based filter could not extract this mode properly since the influence of other modes could not be eliminated. Thus, the FFT-based filter should be applied here.

When using phasor data of two measurement sites, phasor oscillations can be observed clearly through calculating the phase difference of two sites by setting one site as the reference of the phase angle. However, when using measurement data of only one site, it is difficult to define the reference; therefore, the simplified model (3.16) cannot be applied to identify the low-frequency oscillations. Here, Fig. 3.9 shows the low-frequency oscillations extracted by the FFT-based filter using frequency deviation data measured at Miyazaki, Nagoya, and Tokushima, which are located at both ends and the middle part of the system, respectively. The low-frequency oscillations of Miyazaki and Nagoya change in opposite phases. On the other hand, the waveform of Tokushima has a smaller amplitude than the other two sites. This result implies that the fixed node of the low-frequency oscillation, that is $\delta_s = 0$, should exist near Tokushima. Thus, the oscillation model identified by using the information of only one measurement site can be considered by setting this point as the reference of the phase angle.

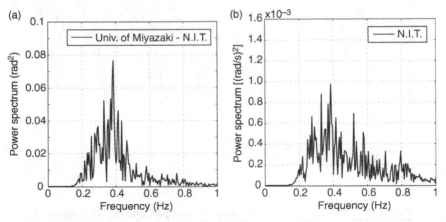

Figure 3.8. FFT spectrum for: (a) phase difference and (b) frequency deviation.

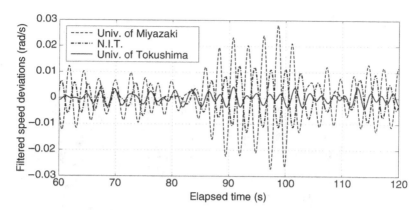

Figure 3.9. Filtered speed deviations of each point.

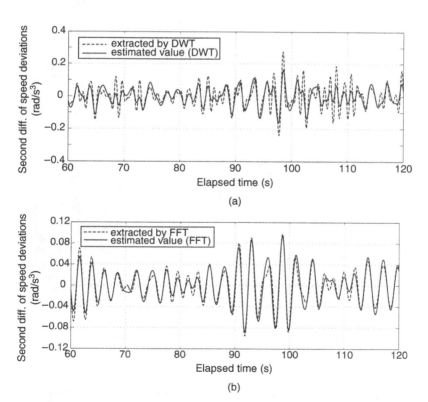

Figure 3.10. Comparison of modeling accuracy: (a) DWT and (b) FFT.

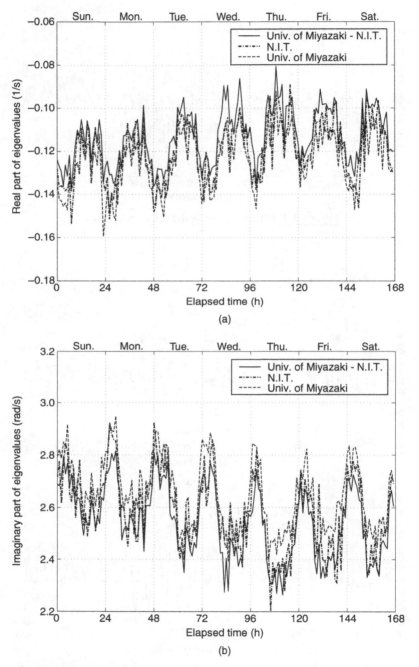

Figure 3.11. Comparison of eigenvalues: (a) real part and (b) imaginary part.

Assuming $\dot{\delta}_s = 0$, the differentiation of the model (3.16) is

$$\begin{bmatrix} \ddot{x}_1 \\ \ddot{x}_2 \end{bmatrix} = \begin{bmatrix} a_1 & a_2 \\ 1 & 0 \end{bmatrix} \begin{bmatrix} \dot{x}_1 \\ \dot{x}_2 \end{bmatrix} \tag{3.19}$$

where $\dot{x}_1 = \delta_1$ and $\dot{x}_2 = \dot{\delta}_1$. The model (3.19) represents the oscillation dynamics of the low-frequency oscillation using the information of frequency deviations.

Figure 3.10 shows the result of a similar investigation as shown in Fig. 3.6. The sum of square errors for DWT and FFT are 110.4 and 69.0, respectively, where the amplitude is normalized by the maximum value. The accuracy of the identified model is improved by applying the FFT-based filtering method. Figure 3.11 shows the comparison results for the stability assessment for 1 week from August 13 to 19, 2006:

1. The eigenvalues estimated by (3.16) identified using the phase difference between Miyazaki and Nagoya.
2. The eigenvalues estimated by (3.19) identified using the frequency deviations of Nagoya.
3. The eigenvalues estimated by (3.19) identified using the frequency deviations of Miyazaki.

The sum of square errors for Miyazaki–Nagoya, Nagoya, and Miyazaki are 32.0, 26.9, and 21.8, respectively. Therefore, the accuracy of the identified mode does not deteriorate even when using the frequency deviation data-based method. Since the eigenvalues can be estimated successfully, the frequency deviation data are effective in evaluating the stability of low-frequency oscillations.

3.4 SUMMARY

This chapter describes the small-signal stability assessment with phasor measurements. Particularly, the stability of the interarea low-frequency oscillation mode has been investigated by adopting the method to identify the oscillation dynamics with a simple oscillation model. The filtering approach improves the accuracy of the estimated eigenvalues. The stability can be evaluated successfully by the presented approach.

REFERENCES

1. J. A. Demcko, S. Pillutla, and A. Keyhani, Measurement of synchronous generator data from digital fault recorders for tracking of parameters and field degradation detection, *Electr. Power Syst. Res.*, **39**, 205–213, 1996.
2. C. Rehtanz and D. Westermann, Wide area measurement and control system for increasing transmission capacity in deregulated energy markets. In: *Proceedings of the 14th Power Systems Computation Conference*, 2002.

3. R. Avila-Rosales and J. Giri, Wide-area monitoring and control for power system grid security. In: *Proceedings of the 15th Power Systems Computation Conference*, 2005.

4. N. Kakimoto, M. Sugumi, T. Makino, and K. Tomiyama, Monitoring of interarea oscillation mode by synchronized phasor measurement, *IEEE Trans. Power Syst.*, **21**(1), 260–268, 2006.

5. A. G. Phadke, Synchronized phasor measurements in power systems, *IEEE Comput. Appl. Power*, **6**(2), 10–15, 1993.

6. A. G. Phadke, et al., The wide world of wide-area measurement, *IEEE Power Energy Mag.*, **6**(5), 52–65, 2008.

7. P. M. Anderson and A. A. Fouad, *Power System Control and Stability*, The Iowa State University Press, Ames, IA, 1977.

8. I. J. Peez-Arriaga, G. C. Verghese, and F. C. Schweppe, Selective modal analysis with application to electric power systems, part 1: heuristic introduction, *IEEE Trans. Power Apparatus Syst.*, **101**(9), 3117–3125, 1982.

9. T. Hashiguchi, M. Watanabe, A. Matsushita, Y. Mitani, O. Saeki, K. Tsuji, M. Hojo, and H. Ukai, Identification of characterization factor for power system oscillation based on multiple synchronized phasor measurements, *Electr. Eng. Jpn.*, **163**(3), 10–18, 2008.

10. The MathWorks, *MATLAB Wavelet Toolbox. Wavelet Toolbox User's Guide*, 2002.

11. Technical Committee of IEEJ, *Japanese Power System Models*, 1999. Available at http://www2.iee.or.jp/ver2/pes/23-st_model/english/index.html.

12. R. Tsukui, P. Beaumont, T. Tanaka, and K. Sekiguchi, Intranet-based protection and control, *IEEE Comput. Appl. Power*, **14**(2), 14–17, 2001.

4

GRAPHICAL TOOLS FOR STABILITY AND SECURITY ASSESSMENT

Power system dynamic analysis and control synthesis is becoming more significant today due to the increasing size, changing structure, emerging renewable energy sources and new uncertainties, environmental constraints, and the complexity of modern power systems. In response to this crudity, the present chapter introduces new analytic graphical tools for studying the power system dynamics, stability evaluation, and security assessment.

These graphical tools could be useful in developing new control schemes in normal, predictive, and emergency states. The proposed graphical tools employ well-known power system indices, that is, frequency, voltage, and angle deviations in proper multidimensional planes. The presented planes are angle–voltage deviations, voltage–frequency deviations, frequency–angle deviation, and wave propagation graphs.

4.1 IMPORTANCE OF GRAPHICAL TOOLS IN WAMS

The power system is the largest man-made system with numerous unpredictable events and disturbances that may often be characterized by instability phenomena, leading to the triggering of component protections, resulting in a widespread interruption or blackout,

Power System Monitoring and Control, First Edition. Hassan Bevrani, Masayuki Watanabe, and Yasunori Mitani.
© 2014 John Wiley & Sons, Inc. Published 2014 by John Wiley & Sons, Inc.

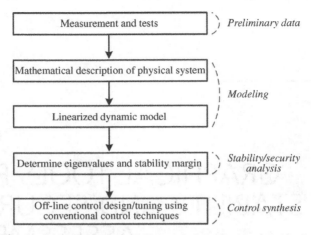

Figure 4.1. Traditional stability analysis and control synthesis.

where it is difficult to anticipate the affected region. Instability phenomena related to severe events are mainly large-disturbance voltage, rotor angle, and frequency instability [1].

Due to the growing size and complexity of power systems, monitoring-based and graphical analysis tools are increasingly used in modern power systems worldwide, providing new approaches for stability/security assessment and control design. A simplified schema for traditional stability analysis and control synthesis is shown in Fig. 4.1.

Simplicity of analysis using graphical/descriptive tools without considering the dynamic model is the main advantage of these tools in large-scale power systems. The improved situational awareness provided by PMUs and wide-area monitoring systems (WAMSs) opens further improvements of graphical analysis, identification, and control tuning methods, providing robustness against unforeseen disturbances. Figure 4.2 shows

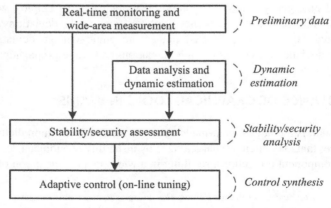

Figure 4.2. Modern stability analysis and control synthesis.

a new scheme for stability analysis and control synthesis using wide-area monitoring-based descriptive methods. The main purpose of these tools is easy monitoring to prevent instability and to maintain an interconnected operation of the power system.

4.2 ANGLE–VOLTAGE DEVIATION GRAPH

As shown in Chapter 7, phase difference–voltage deviation ($\Delta \delta - \Delta V$) graphs can be employed to provide a new useful design tool for the coordination of automatic voltage regulator (AVR) and power system stabilizer (PSS) systems. The conflict behavior of voltage and angle indices makes the proposed graph a powerful tool for voltage regulation and small-signal stability studies [2].

After any load/generation disturbance, a graph related to each connected generator can be plotted in the plane of phase difference versus voltage deviation ($\Delta \delta - \Delta V$), which is defined as follows [2]:

$$\Delta \delta_i(t) = \delta_i(t) - \delta_{0i} \tag{4.1}$$

$$\Delta V_i(t) = V_i(t) - V_{0i} \tag{4.2}$$

where δ_{0i}, V_{0i} are the initial values of rotor angle and terminal voltage related to generator i, respectively.

In order to obtain a new criterion that is independent of the fault type, parameter normalization is required. This procedure can be done via the following terms:

$$\Delta \delta_{\max}(t) = \max\{|\Delta \delta_i(t)|\} \tag{4.3}$$

$$\Delta V_{\max}(t) = \max\{|\Delta V_i(t)|\} \tag{4.4}$$

$$\Delta \delta_i = \frac{\Delta \delta_i(t)}{\Delta \delta_{\max}(t)} \tag{4.5}$$

$$\Delta V_i = \frac{\Delta V_i(t)}{\Delta V_{\max}(t)} \tag{4.6}$$

Economic reasons and environmental constraints have led to the transmission lines of systems operating at the maximum transfer power capacity. Therefore, the desired operation of a power system after disturbances could be achieved when the system returns to the planned operating level characterized by voltage profile, nominal frequency, and power flow configuration. The mathematical representation of power flow could be shown by

$$P_{ij} = \frac{|V_{0i} + \Delta V_i(t)||V_{0j} + \Delta V_j(t)|}{X_{ij}} \sin\left[(\delta_{0i} + \Delta \delta_i(t)) - (\delta_{0j} + \Delta \delta_j(t))\right] \tag{4.7}$$

In order to achieve a stable condition and desired performance, the following conditions must be simultaneously met:

1. Voltage deviations for each generator converge to zero.
2. All rotor angle deviations converge to the same value.

In ideal situations, all connected generators completely satisfy the above conditions. Therefore, each generator will be fixed at the (1,1) point in the normalized plane of phase difference versus voltage deviation after a fault. However, after a fault in a practical system, the system returns to a new operating point and hence, the voltage and rotor angle deviations of generators differ from each other. Therefore, control actions have to be done in such a way that they conduct each generator terminal voltage deviations to zero (0 in the normalized voltage deviation axis) as well as the rotor angle deviations to the same value, that is, 1 in the normalized phase difference axis. In other words, for stable, secure, and reliable system operation, all generator operating points must be located in the minimum distance from the (1,0) point in the normalized $(\Delta\delta - \Delta V)$ plane. This condition could be mathematically represented by the following minimization problem:

$$\min\{\|\Delta\delta_i - \Delta V_i\| - \|(1,0)\|\} \tag{4.8}$$

or

$$\min\{\sqrt{(\Delta\delta_i)^2 + (\Delta V_i)^2} - 1\} \tag{4.9}$$

The optimum value can be achieved when derivation of the above equation becomes zero:

$$\frac{2\Delta\delta_i + 2\Delta V_i[(\mathrm{d}(\Delta V_i))/(\mathrm{d}(\Delta\delta_i))]}{2\sqrt{(\Delta\delta_i)^2 + (\Delta V_i)^2}} = 0 \tag{4.10}$$

This equation leads to

$$2\Delta\delta_i + 2\Delta V_i \frac{\mathrm{d}(\Delta V_i)}{\mathrm{d}(\Delta\delta_i)} = 0 \tag{4.11}$$

which gives

$$\Delta\delta_i = \pm\Delta V_i \quad \text{or} \quad |\Delta\delta| = |\Delta V| = 0.707 \tag{4.12}$$

As previously stated, control actions try to return all the connected generators to the prefault normal condition, that is, the (1, 1) point in the normalized plane. Therefore,

$$\|\Delta\delta_i\| \geq 0.707; \quad \|\Delta V_i\| \geq 0.707 \tag{4.13}$$

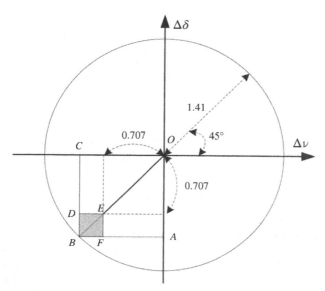

Figure 4.3. Stable region of power system.

On other hand, according to parameter normalization, one can write

$$\|\Delta\delta_i\| \leq 1; \ \|\Delta V_i\| \leq 1 \tag{4.14}$$

It is noteworthy that Equation (4.12) demonstrates the generic desired borders for each probable condition. The occurred fault in a system specifies the sign of phase and voltage variations and thereafter the required borders, that is, the sign of Equation (4.12), to control the system.

From Equations (4.13) and (4.14), the stable region of the power system in the normalized plane of phase difference versus voltage deviation can be represented as shown in Fig. 4.3. It is assumed that the AVR and PSS have a dead band in response, which could be represented in the BDEF area. For the generators located in the BDEF area, the AVR and PSS are well tuned. However, for the other regions, the AVR and PSS parameters should be retuned to return the operating point to the BDEF area. After installing the AVR and PSS in a power system, usually the appropriate tunable parameters are their gains. In Chapter 7, an algorithm is introduced for proper retuning of the AVR and PSS gains. Based on Equation (4.13), the operation region of the controlled systems on the proposed graph could be mathematically explained by

$$\|\Delta\delta_i + \Delta V_i\| \geq 0.707 \tag{4.15}$$

In an interconnected power system, in order to provide the required power in each area, deviations of voltages from nominal values must be compensated by the phase differences of existing generators. However, increasing generator phase differences, that is, decreasing voltages profile, may lead a system to an unstable mode. In this case, another control option such as the static VAR compensator (SVC) must be employed to avoid power system instability. The SVC is implemented to compensate the required reactive power in order to avoid undesirable phase differences increment. Therefore, Equation (4.15) reveals that in the presence of only AVR and PSS, a generator must be located at the minimum distance of 0.707. This statement could be mathematically explained by

$$\min\{\|\Delta\delta_i + \Delta V_i\| - 0.707\} \tag{4.16}$$

Following the same procedure as that used for Equation (4.8), the minimization problem results in $|\Delta\delta| = |\Delta V| = 0.5$. Therefore, the operating region of AVR and PSS can be described as follows:

$$\|\Delta V_i\| \geq 0.5; \|\Delta\delta_i\| \geq 0.5 \tag{4.17}$$

As mentioned, for the remaining regions in addition to the AVR and PSS, the SVC may be also required to enhance the system performance. This problem can be clearly seen in power systems with wind power penetration. Wind generators usually employ induction machines rather than synchronous ones. Therefore, the required reactive power for initializing the induction machines is taken from the grid and hence, the bus voltages are decreased. This issue is considered for the wide-area AVR–PSS–SVC coordinator synthesis in Chapter 7.

4.3 SIMULATION RESULTS

The capability of the introduced graphical tool is investigated after application on several large-scale interconnected power systems. Here, this issue is illustrated using the New York/New England system, a 16-machine test system as shown in Fig. 4.4. All

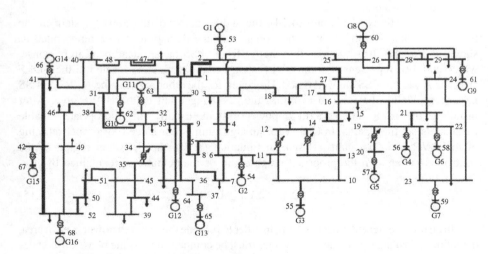

Figure 4.4. Single line diagram of IEEE 68-bus test system [3].

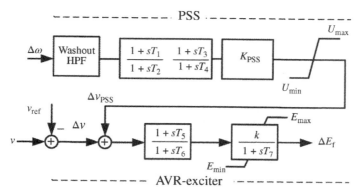

Figure 4.5. Excitation system and PSS structure.

the generators are represented by a two-axis model and equipped with PSS. Structures of the implemented controllers, that is, AVR and PSS are also shown in Fig. 4.5. System parameters are given in Ref. [3], and some simulation parameters are rewritten in Appendix A.

4.3.1 Disturbance in Generation Side

As the first test scenario, G1 has been tripped and the response in the voltage and angle point of view is studied. Figure 4.6a shows the time response of voltage and angle of all available generators within 20 s after outage of G1. The behavior of the system in the $\Delta\delta - \Delta v$ plane is also shown in Fig. 4.6b. In the performed test scenario, the overall generated power is decreased, while the system load is fixed; therefore, the generator angles and bus voltages are decreased. The amount of reduction in rotor angles and bus voltages depends on the system inertia and active/reactive power of the tripped generator.

It can be seen that the system remains stable using the existing control units. The amount of decrease in rotor angles of all generators following the fault are almost the same, and all bus voltages are going to zero without significant oscillations.

Normalized $\Delta\delta - \Delta v$ parameter distribution based on the introduced criterion at 1, 10, and 20 s, following the above-mentioned fault, is shown in Fig. 4.7. It is shown that except for the first sample (at 1 s), in other instants the generators remained in the stable region.

As another test scenario, the system behavior is studied following the outage of G2. The results are shown in Figs 4.8 and 4.9. In this case also, the power system is capable of maintaining stability using the existing control loops.

However, the outage of G12, as shown in Fig. 4.10, presents a different condition. In this case, the system encounters transient instability, that is, first swing instability. The capability of the introduced criterion for moments before instability is also shown in Fig. 4.11. Divergence of the voltage deviations and no similarity in rotor angle deviation clearly depict system instability.

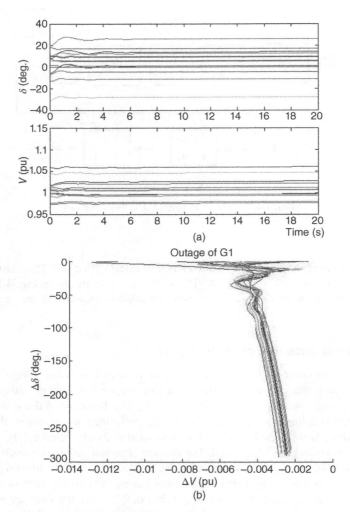

Figure 4.6. System response for outage of G1: (a) time-domain results and (b) $\Delta\delta - \Delta v$ graph.

Figure 4.11 shows that only a few generators (G14, G15, and G16) remain in the stable region and other units are outside of it. So the system is not capable of holding the overall stability. To solve this problem, effective coordination between the PSS and AVR (as well as other control devices such as SVC) can be considered as a suitable alternative. Some coordination approaches are explained in Chapter 7.

4.3.2 Disturbance in Demand Side

All the above-mentioned test scenarios are considered only in the generation side (generator outage). Here, to demonstrate the generality of the given criterion, the

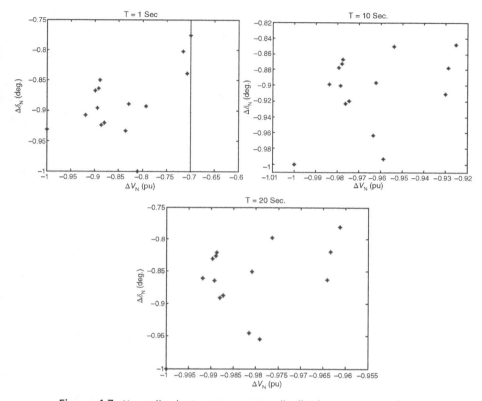

Figure 4.7. Normalized $\Delta\delta - \Delta v$ parameter distribution at 1, 10, and 20 s.

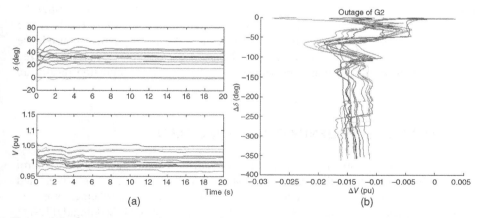

Figure 4.8. System response for outage of G2: (a) time-domain results and (b) $\Delta\delta - \Delta v$ graph.

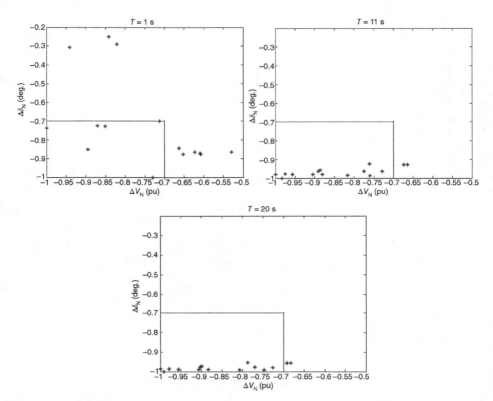

Figure 4.9. Normalized $\Delta\delta - \Delta v$ parameter distribution at 1, 10, and 20 s.

load disturbance is also examined. For this purpose, a large load block, that is, 6000 MW active power and 300 Mvar reactive power, is disconnected from bus 37. To stabilize the system following this severe load disturbance, a coordination algorithm (such as one described in Chapter 7) is needed for the existing control devices. System response in the presence of a coordinator is shown in Figs 4.12 and 4.13, and the relevant normalized $\Delta\delta - \Delta v$ parameter distribution is presented in Fig. 4.14. More simulation results are given in Ref. [4].

4.4 VOLTAGE–FREQUENCY DEVIATION GRAPH

Frequency and voltage are more frequent decision indices in many power system control strategies such as automatic generation control, automatic voltage regulation, and emergency control schemes [5]. For example, most load shedding schemes proposed so far use frequency or voltage information via underfrequency load shedding (UFLS) or undervoltage load shedding (UVLS) algorithms. Furthermore, many protection devices

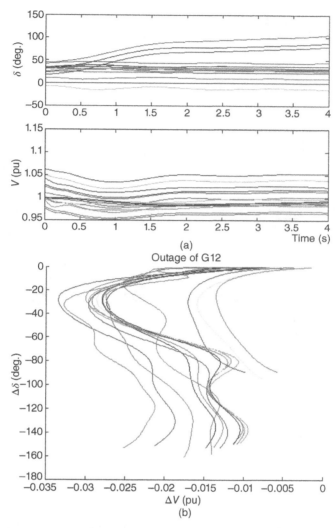

Figure 4.10. System response for outage of G12: (a) time-domain results and (b) $\Delta \delta - \Delta v$ graph.

such as underfrequency and undervoltage relays are usually working in power systems based on these indices.

4.4.1 $\Delta V - \Delta F$ Graph for Contingency Assessment

In this section, a graphical analysis tool to study the system dynamic (frequency and voltage) behavior following a disturbance is introduced. This tool can be effectively used

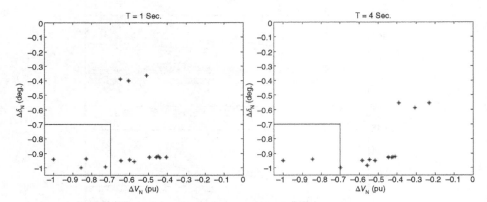

Figure 4.11. Normalized $\Delta\delta - \Delta v$ parameter distribution at 1 and 4 s.

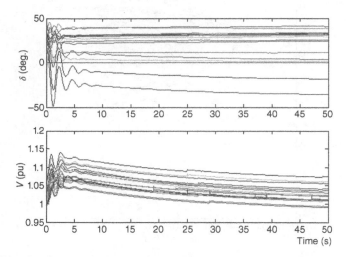

Figure 4.12. System time-domain response for a severe load disturbance.

for the postcontingency stability analysis issue. Following a contingency event, the power system operating point deviates from its stable precontingency state. If the contingency is not severe enough, the system may converge to another stable point. Otherwise, the system state may move to an unstable region.

As mentioned, for emergency control purposes, voltage and frequency are two suitable measurable variables that are used to illustrate the state of the system following an event. Therefore, it seems that the frequency–voltage deviations ($\Delta v - \Delta f$) plane can be used as a useful graphical tool to see the state of the system following a contingency.

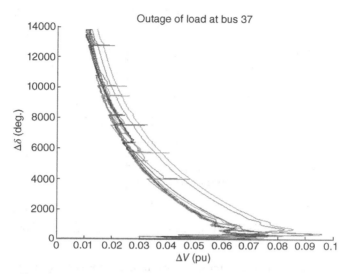

Figure 4.13. The $\Delta\delta - \Delta v$ graph for a severe load disturbance.

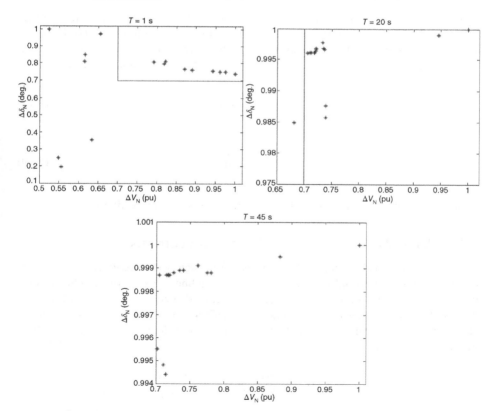

Figure 4.14. Normalized $\Delta\delta - \Delta v$ parameter distribution at 1, 20, and 45 s.

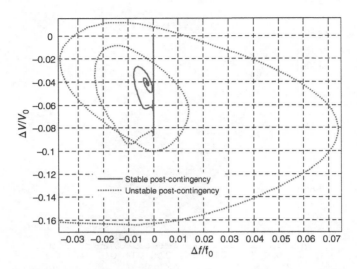

Figure 4.15. Phase trajectory for stable and unstable postcontingency scenarios.

Figure 4.15 shows this two-dimensional plane, to show the system frequency–voltage behavior after a serious disturbance, with two scenarios: unstable trajectory (without load shedding) and stabilized trajectory using a load shedding plan. The states that are used in this trajectory are Δf and ΔV, represented in the following complex statement [5]:

$$S = \hat{\Delta f} + j \hat{\Delta v} \tag{4.18}$$

where

$$\hat{\Delta f} = \frac{\Delta f}{f_0}, \quad \hat{\Delta v} = \frac{\Delta v}{V_0} \tag{4.19}$$

f_0 and V_0 are the normal values of frequency and voltage, respectively.

4.4.2 $\Delta V - \Delta F$ Graph for Load Shedding Synthesis

To design a new load shedding algorithm based on state variables given in (4.19), the two-dimensional threshold boundaries should be defined instead of one-dimensional threshold values that are used in the conventional load shedding algorithms. Since the threshold movement size in two directions (the UFLS and the UVLS) are not equal, the elliptical shape for the existing boundaries can be considered as a suitable solution. Therefore, assuming

$$x^t = \hat{\Delta f} = \frac{f_t - f_0}{f_0}, y^t = \hat{\Delta v} = \frac{v_t - V_0}{V_0} \tag{4.20}$$

the load shedding control signals (LSCS) can be defined as follows:

$$\text{LSCS}_i = \left(\frac{x^t}{a_i}\right)^2 + \left(\frac{y^t}{b_i}\right)^2 \tag{4.21}$$

where x^t and y^t are new state variables. f_t and V_t are the frequency and voltage at time t, respectively. The a_i and b_i parameters must be computed by designer. A threshold boundary satisfies the inequality $\text{LSCS}_i > 1$.

Considering a standard load shedding strategy such as the one introduced by the Florida Reliability Coordinating Council (FERC) [6], the load shedding steps a_i and b_i parameters can be calculated. In this case, by fixing variable y^t at zero, the proposed algorithm is reduced to the FRCC's UFLS program. Each step is determined by an ellipse and when the phase trajectory reaches each ellipse, the corresponding load shedding step will be triggered. Figure 4.16 shows the result of applied load shedding steps on the $\Delta v - \Delta f$ trajectory plane in the case of G2 outage for an updated version of IEEE 9-bus test system (with wind turbine generators) [7].

The loads are selected to be triggered at the zones where their LSCS parameters (4.21) lead to a load shedding step triggering. When a disturbance happens, the voltage and frequency of a zone will change faster than the others; therefore, the loads in this zone will find the highest priority for load shedding. In order to examine the capability of the proposed graphical tool for the introduced UFVLS scheme, several simulation results on severe load and generation disturbance for IEEE 9- and 24-bus reliability systems are given in Chapter 8.

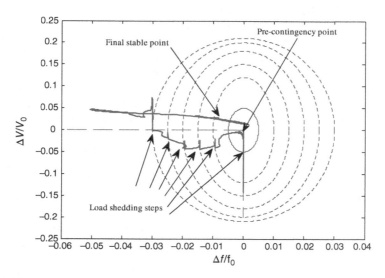

Figure 4.16. The $\Delta v - \Delta f$ graph showing load shedding result following loss of G2 of a nine-bus test system.

4.5 FREQUENCY–ANGLE DEVIATION GRAPH [8]

This section investigates the dynamic characteristics of frequency deviations after a disturbance based on PMU measurements, observing the frequency deviations and phase differences at different points. The power flow behaviors and frequency response following disturbances are studied using the relation between phase differences and frequency deviations (on a plane of frequency deviation versus phase difference) as well as wavelet analysis.

The varying phase angle and phase difference between two points may provide useful information. Increasing phase angle represents that the measured frequency in the system is higher than the nominal frequency (e.g., 60 Hz). Decreasing phase angle means the measured frequency is lower than nominal frequency. The time derivative of phase angle corresponds to the deviations of system frequency. Therefore, the frequency deviations can be calculated as

$$\Delta f_n = \frac{\delta_{n+1} - \delta_n}{360 \Delta t_n} \tag{4.22}$$

where Δt_n [s] is a sampling interval of sequential phase data δn and n is the number of accumulated phase angle data. Therefore, frequency variation can be observed by the PMU with the accumulation of the sequential frequency deviations Δf_n. The variation of phase difference between points A and B represents the power flow deviation between A and B. Increase of phase difference shows the increase of power flow and vice versa.

The PMU corrects its internal clock by the time stamps of the global positioning system (GPS). The clock time error is guaranteed within 1 μs and phase angle error is guaranteed to be less than 0.1°. Therefore, the accuracy of time and phase angle is enough to analyze the frequency oscillations of the whole power system, concerning the secondary frequency control time scale. It is shown in Ref. [9] that the probability density $g(\Delta P_{\text{tie}}, \Delta f)$ of two areas has a shape of concentric circle or ellipse when distributions of the area control error signals are of normal type.

The method described is useful for considering frequency response characteristics of an interconnected power system. For this study, the actual value of frequency deviation Δf and the tie-line power flow deviation ΔP_{tie} may be required. Although the frequency deviation Δf can be calculated by the PMU, the tie-line power flow deviation ΔP_{tie} cannot be directly obtained. However, since the phase difference and power flow change between two points have a close relationship, to indicate the power flow variation, the phase difference between two points is calculated from the PMU measurements.

For this purpose, the $\Delta f - \Delta \delta$ plot could be useful. The vertical axis is considered for frequency deviation Δf, and the horizontal axis is used for phase difference between two points ($\delta_{AB} = \delta_A - \delta_B$). Two examples are shown in Figs 4.17a and 4.18a. The data are related to a disturbance in Japan for Osaka University (O), Hiroshima University (H), and the University of Tokushima (T). The plots present almost a shape of a concentric ellipse. The frequency distributions of plotted data (by 0.04° steps of δ) are also displayed in Figs 4.17b and 4.18b, respectively. Both histograms show normal distributions, approximately.

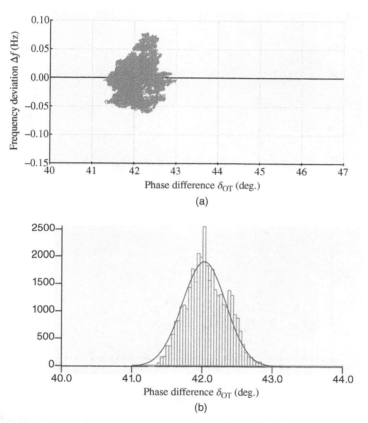

Figure 4.17. (a) $\Delta f - \Delta \delta$ plot between Osaka University and the University of Tokushima. (b) Histogram of the plot.

The histograms can provide an approximated steady-state value of phase difference between two points. In the case of the phase difference between Hiroshima University and the University of Tokushima, the initial value $\Delta \delta_{T0}$ can be estimated as 104.3°. In addition to this, the stiffness of the two tie-lines can be also guessed by comparing both histograms.

Because the power system has a high-voltage DC (HVDC) transmission line between Kansai and Shikoku, the phase difference does not directly indicate the power flow between them. However, it can be said by the histogram that the interconnection between Kansai and Shikoku is very tight.

Figure 4.19a shows an ellipse on the left side of the plot, and some data are also expanded toward the bottom right of the plane that is redrawn in Fig. 4.19b. The extracted data shown in Fig. 4.19b are fitted with the following line equation:

$$\Delta f = -0.00184\delta_{HT} + 1.891 \tag{4.23}$$

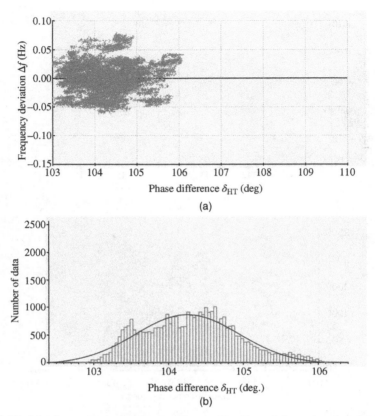

Figure 4.18. (a) $\Delta f - \Delta \delta$ plot between Hiroshima University and the University of Tokushima. (b) Histogram of the plot.

Equation (4.23) presents a linear relationship for frequency deviation and phase angle difference. When the unbalanced power occurs in one system, frequency deviation Δf and power flow deviation on a tie-line represents a linear relationship. Because the power flow deviation is closely related to the phase angle difference between two areas, frequency deviation and phase angle difference are expected to show the same linear relation.

4.6 ELECTROMECHANICAL WAVE PROPAGATION GRAPH

When a disturbance takes place in a power system, the rotor angle of the generators near where the fault occurred deviates from the base frames. This deviation propagates through the power system. Hence, for large disturbances, this may lead to catastrophic outages, and finally a blackout.

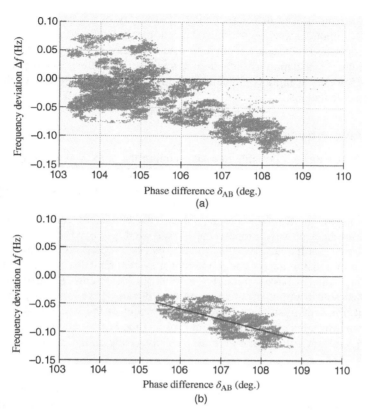

Figure 4.19. (a) $\Delta f - \Delta \delta$ plot between Hiroshima University and the University of Tokushima. (b) Histogram of the plot.

Because of deregulation and economic issues in power system management and the rapid growth of the number of electrical power consumers, transmission lines are usually working close to their stability limits. In other words, nowadays power systems are working under stress. One of the problems that a stressed power system encounters is the prediction of instability location. The occurrence of a disturbance at a certain point in large-scale power systems may initiate instability in another point of the grid due to the electrical wave propagation phenomenon.

The power system can be considered as a continuum, and the propagated electromechanical waves initiated by faults and other random events can be studied to provide a new scheme for stability investigation of a large dimensional system. For this purpose, the measured electrical indices (such as rotor angle and bus voltage) following a fault in different points on the network are used, and the behavior of the propagated waves through the lines, nodes, and buses is analyzed. The impact of weak transmission links on a progressive electromechanical wave using the energy function concept is addressed.

It is also emphasized that determining the severity of a disturbance/contingency accurately, without considering the related electromechanical waves, hidden dynamics, and their properties is not secure enough. Considering these phenomena takes heavy and time-consuming calculation, which is not suitable for online stability assessment problems. However, using a continuum model for a power system reduces the burden of complex calculations.

4.6.1 Wave Propagation

Emerging PMUs make it possible to easily monitor the angle wave propagation in the power systems. They use GPS to provide a synchronous time in the power systems, which may be distributed among the continents. There are many research reports about wave traveling phenomenon in a power system as a continuum model. Using a continuum model for a power system reduces the burden of complex calculations. The electromechanical wave propagation issue is well discussed in Refs [10–12].

It is shown that like a wave, disturbances can be propagated through the system. Propagation of a disturbance through the power system may threaten the stability if it becomes large or if it passes through the weak lines/elements. Furthermore, it may cause undesirable tripping of some protection devices [12]. To study the electromechanical transients, usually a large detailed model of the whole power system is needed. Preparing such system models and solving the heavy equations are very time-consuming, and often it will not give us a sensible view on the global power system stability.

Studying a system based on wave propagation analysis is needed to observe an overall view of the system. Therefore, a descriptive approach could be considered as a suitable alternative to analyze the power system based on wave propagation. All studies performed on wave propagation over the years can be categorized into two groups: voltage and electromechanical wave propagation. Electromechanical wave propagation is produced when a rotor angle of a synchronous generator deviates from its base frame following a disturbance.

In this section, the measured electrical indices (such as the rotor angle and the bus voltage) following a fault in different points among the network are used, and the behavior of the propagated waves through the lines, nodes, and buses is analyzed. It is shown that the reason for the instability and abnormal operations can be found from the analysis of electrical wave propagation phenomenon and identifying the weak links in the system. The impact of weak transmission links on a progressive electromechanical wave using the energy function concept is addressed. Some parts of this section are expanded in Ref. [13].

4.6.1.1 Voltage Wave Propagation. The study on power system stability can be divided into two general groups: static stability and transient stability. In static stability studies, the goal is to verify system stability when encountering low-amplitude slow disturbances. However, in the transient stability problems, a large and sudden disturbance occurs. Dynamic stability is an improved form of static stability in the case of low-amplitude disturbances with longer lifetime [14].

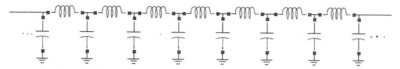

Figure 4.20. Distributed model of a transmission line.

In static stability studies, the wavelength is about 6000 km for 50 Hz [15]. Therefore, in these studies a lumped model of a line is adequate. However, for the transient stability problems where higher frequencies exist, the traveling waves cannot be ignored. In other words, in this situation when a signal exists at one end of a line, there is no guarantee that it will appear at the other end of the line at the same moment. Indeed, the wave travels through the lines with a delay. This delay is actually because of charging of the capacitance and inductance elements of the line [15]. In the lumped π-model, only two capacitors and one inductor are used; and in the T-model there are only two inductors and one capacitor. Therefore, the voltage variations are immediately sensed at the other end of the line. However, as shown in Fig. 4.20, a distributed model contains a lot of elements where each element includes one inductor and one capacitor. When a voltage source is applied at one end of the line, the first capacitor is immediately charged; however, because of the first inductor, the second capacitor charges when the inductor is charged, and this causes a delay. This delay also exists for the next elements.

One of the most important parameters in this issue is the *characteristic impedance*, which is calculated by

$$Z = \sqrt{\frac{L}{C}} \tag{4.24}$$

where L and C are defined as line inductance and capacitance values per length of line, respectively.

When a voltage wave passes through a point in which the characteristic impedance is changed, the magnitude of the reflected wave and the magnitude of the wave that gets through are dependent on the value of the characteristic impedances of two lines (for example when an overhead transmission line is connected to an underground cable).

The wave *propagation velocity* is another important parameter that can be calculated as follows:

$$v = \frac{1}{\sqrt{LC}} \tag{4.25}$$

4.6.1.2 Rotor Angle Wave Propagation. As discussed, the angle wave propagation, which is called electromechanical wave propagation, starts when a generator rotor deviates from its base frame, then propagates throughout the power system. The velocity of angle wave propagation is slower than voltage wave propagation [10,12].

As was argued in Ref. [12], the nature of this phenomenon is not completely known. Since the waves may lead to loss of synchronism for stressed power systems, studying them is one of the interesting issues in power system stability analysis and security assessment.

The formulation of wave motion as nonlinear partial differential equations in a two-dimensional surface is introduced in Ref. [10]. This formulation is in the form of wave equations. The swing equation of a generating unit can be expressed as follows:

$$\frac{2H}{\omega}\ddot{\delta} + \omega D\dot{\delta} = P_a = P_m - P_e \tag{4.26}$$

where H, ω, D, δ, P_e, and P_m are inertia constant, angular speed, damping factor, rotor angle, electrical output power, and mechanical input power, respectively.

The swing equation is introduced as a second-order hyperbolic wave equation in Ref. [10]. Using (4.26), lets us introduce a power system as a continuum system. The methodology of extracting the continuum model for a 2D power system is well established in Ref. [11]. For a mesh network power system wherein the generators and loads are located at distributed discrete points, introducing a 2D continuum model needs the introduction of system parameters in the form of smooth functions. To distribute the parameters, a *Gaussian filter*, which is the most common smoothing tool, can be suggested [11].

4.6.2   Angle Wave and System Configuration

Assume a highly stressed large-scale power system where its elements are working near their stability margins. In this case, disturbances may easily force the system into cascading failures and even a blackout.

On the other hand, following a contingency, the voltage/angle deviations propagate through the power system. These traveling waves may pass through the highly stressed elements and trigger an instability condition. By using the wave propagation phenomenon, a descriptive approach for power system stability assessment can be proposed. This descriptive tool can be used to track the system dynamic behaviors, to validate the system performance against likelihood events, and to predict the next stable operating point. Here, the main aim is to look for an approach to identify a proper emergency action immediately.

Following dangerous events, there are some choices for emergency control actions. Sometimes a load shedding scheme or a generation rescheduling action can maintain system stability. However, for serious disturbances, it may be needed to separate the system into two or more islands with different control systems.

In transient stability studies, there are some approaches for system stability assessment following a contingency. For example *Equal area criterion* is one of these approaches, which is only used for one machine connected to an infinite bus or for two-machine systems [14]. The energy function criterion for online stability assessment of a large power system is used in Ref. [16]. The proposed criteria show that a system can stay stable if the generators can release their complete energy value following a contingency.

4.6.2.1 Ring Systems.

If the number of elements in a power system is large with a set of distributed generator parameters, simulation results for a discretized model will be close to the continuum one [10]. The following equations can be used for modeling an N-bus ring power system:

$$\ddot{\delta}_k + D\dot{\delta}_k = P_m^k - [2 - \cos(\delta_k - \delta_{k-1}) - \cos(\delta_k - \delta_{k+1})] - b[\sin(\delta_k - \delta_{k-1}) + \sin(\delta_k - \delta_{k+1})],$$

for $k = 2, 3, ..., N-1$

$$\ddot{\delta}_1 + D\dot{\delta}_1 = P_m^1 - [2 - \cos(\delta_1 - \delta_N) - \cos(\delta_1 - \delta_2)] - b[\sin(\delta_1 - \delta_N) + \sin(\delta_1 - \delta_2)]$$

$$\ddot{\delta}_N + D\dot{\delta}_N = P_m^N - [2 - \cos(\delta_N - \delta_{N-1}) - \cos(\delta_N - \delta_1)] - b[\sin(\delta_N - \delta_{N-1}) + \sin(\delta_N - \delta_1)]$$

$$(4.27)$$

The value of P_m^k can be achieved from steady-state values of angles ($\ddot{\delta}_k = \dot{\delta}_k = 0$). The steady-state values of angles for an N-bus ring system can be calculated as

$$\delta_k^{eq} = \frac{2\pi k}{N} \tag{4.28}$$

Now, consider the 64-bus ring power system given in Ref. [10]. A Gaussian disturbance around line 15–16 can be implemented as follows:

$$\delta_k = \delta_k^{eq} + \frac{1}{2}e^{-0.1(k-15.5)^2} \tag{4.29}$$

The simulation results, which are regeneration of the given example in Ref. [10], are illustrated in Fig. 4.21. In Fig. 4.21a, the wave propagation is illustrated versus time, while in Fig. 4.21b, the angle of wave is plotted versus the bus number for different time slots. As can be seen, when a deviation appears in the rotor angle of a generator, it propagates and in the traveling path, it may encounter weak links, which may lead to cascading failures [10–13,16].

4.6.2.2 2-D Systems.

Now, consider a meshed power system with a configuration shown in Fig. 4.22, in the Cartesian characteristic. Each point represents a bus and each connection represents a transmission line. For simplicity, assume that each bus consists of a generator, or a load, or nothing. To obtain the necessary equations, assume that a generator with a mechanical power of P_m^A is connected to the point A (Fig. 4.22).

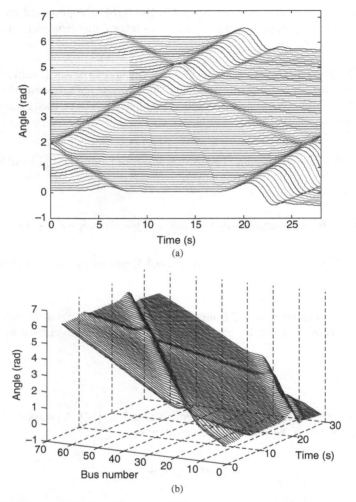

Figure 4.21. Electromechanical wave propagation on 64-bus ring system for angle versus (a) time and (b) bus number.

Considering the connections, P_e^A is the sum of electrical powers transferred from bus A to its neighbor buses 1, 2, 3, and 4. Therefore, P_e^A can be calculated as

$$P_e^A = \sum_{k=1}^{4} P_e^{Ak} + P_L^A \tag{4.30}$$

where

$$P_e^{Ak} = \frac{V^A V^k}{x^{Ak}} \sin\delta^{Ak}, \quad k = 1, 2, 3, 4 \tag{4.31}$$

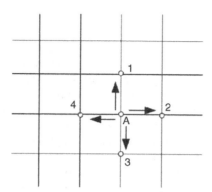

Figure 4.22. Network configuration.

Now, the swing equations can be easily written for each point [13,17]. Using the earlier descriptions, an example is presented in Fig. 4.23, which illustrates the propagation of angle wave through a 2D power system. At $t=0$, a disturbance occurs in the middle of the network then it propagates throughout the system. The system situation following the disturbance is shown in a few time slots.

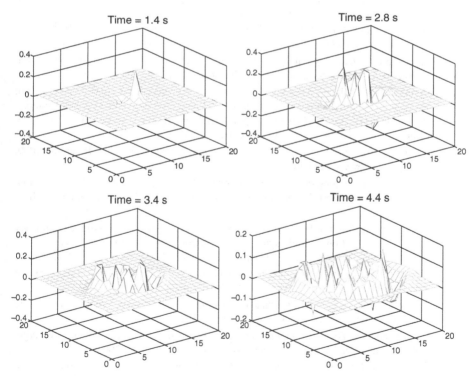

Figure 4.23. Wave propagation through a network in different time slots.

4.7 SUMMARY

Power system stability and security assessment has been considered as an important control issue for secure system operation over many years. Currently, because of expanding physical setups and the penetration of distributed generators and renewable energy sources, the problem has become more significant than in the past.

In the present chapter, for the purpose of power system stability/security analysis and control synthesis, some graphical tools are introduced, that is, angle–voltage deviation, voltage–frequency deviation, frequency–angle deviation, and electromechanical wave propagation graphs. The necessity of using the descriptive and graphical tools rather than pure analytical and mathematical approaches in wide-area power system stability and security issues is explained. Some applications for designing wide-area controllers/coordinators as well as emergency control plans are discussed. More detailed applications are addressed in Chapters 7 and 8.

REFERENCES

1. H. Bevrani, *Robust Power System Frequency Control*, Springer, New York, 2011.
2. H. Golpira, H. Bevrani, and A. H. Naghshbandi, An approach for coordinated automatic voltage regulator power system stabiliser design in large-scale interconnected power systems considering wind power penetration, *IET Gen. Transm. Distrib.*, **6**(1), 39–49, 2012.
3. G. Rogers, *Power System Oscillations*, Kluwer Academic Press, Boston, MA, 1999.
4. H. Golpira, Coordinated design of automatic voltage regulator and power system stabilizer in the presence of RESs, Master thesis, University of Kurdistan, Sanandaj, Iran, 2010.
5. H. Bevrani and A. G. Tikdari, An ANN-based power system emergency control scheme in the presence of high wind power penetration. In: L. F. Wang, C. Singh, and A. Kusiak, editors, *Wind Power Systems: Applications of Computational Intelligence*, Springer Book Series on Green Energy and Technology, Springer, Heidelberg, 2010, pp. 215–254.
6. PRC-006-FRCC-01, FRCC Automatic Underfrequency Load Shedding Program, 2009. Available at https://www.frcc.com/.
7. A. G. Tikdari, Load shedding in the presence of renewable energy sources, M.Sc. dissertation, Department of Electrical and Computer Engineering, University of Kurdistan, Sanandaj, Iran, 2009.
8. M. Hojo, K. Ohnishi, T. Ohnishi, Y. Mitani, O. Saeki, and H. Ukai, Analysis of load frequency control dynamics based on multiple synchronized phasor measurements. In: *Proceedings of the 15th Power Systems Computation Conference (PSCC), Liege, August 22–26*, 2005.
9. T. Sasaki and K. Enomoto, Study of generation control and generation control performance standards, *Trans. IEE Jpn.*, **121B**(3), 307–318, 2001 (in Japanese).
10. J. S. Thorp, C. E. Seyler, and A. G. Phadke, Electromechanical wave propagation in large electric power systems, *IEEE Trans. Circuits Syst.*, **45**(6), 614–622, 1998.
11. M. Parashar and J. S. Thorp, Continuum modeling of electromechanical dynamics in large-scale power systems, *IEEE Trans. Circuits Syst.*, **51**(9), 1851–1858, 2004.
12. A. G. Phadke and J. S. Thorp, *Synchronized Phasor Measurements and Their Applications*, Springer, New York, 2008.

13. A. G. Tikdari, H. Bevrani, and G. Ledwich, A descriptive approach for power system stability and security assessment. In: P. Vasant, N. Barsoum, and J. Webb, editors, *Innovation in Power, Control and Optimization: Emerging Energy Technologies*, IGI Global, Hershey, PA, 2011.

14. P. Kundur, *Power System Stability and Control*, McGraw-Hill, New York, 1994.

15. L. V. D. Sluis, *Travelling Waves, Transients in Power Systems*, Wiley, 2001, pp. 31–56.

16. K. R. Padiyar and S. Krishna, Online detection of loss of synchronism using energy function criterion, *IEEE Trans. Power Deliv.*, **21**(1), 46–55, 2006.

17. H. Bevrani and A. G. Tikdari, Power system stability analysis based on descriptive study of electrical indices. In: *ASIJ 5th Conference, Tokyo, March 6*, 2010.

5

POWER SYSTEM CONTROL: FUNDAMENTALS AND NEW PERSPECTIVES

The term *power systems control* is used to define the application of control theorems and relevant technologies to enhance the power system functions during normal and abnormal operations. Power system control refers to keeping a desired performance and stabilizing power system following various disturbances, such as short circuits and loss of generation and/or load. Power system stability and control was first recognized as an important problem in the 1920s [1,2]. Over the years, numerous modeling/simulation programs, synthesis/analysis methodologies, and protection schemes have been developed. Power system control can take different forms, which are influenced by the type of instability phenomena. A survey on the basics of power system controls, literature, and achievements is given in Refs [3,4].

In this chapter, fundamental concepts and definitions of power system stability and control are emphasized. The role of power system control in preserving system integrity and restoring the normal operation subjected to physical disturbances is described and some challenges, opportunities, and new perspectives concerning the integration of renewable energy options and distributed generators are introduced.

Power System Monitoring and Control, First Edition. Hassan Bevrani, Masayuki Watanabe, and Yasunori Mitani.

5.1 POWER SYSTEM STABILITY AND CONTROL [5]

Power system stability is defined as "the ability of an electric power system, for a given initial operating condition, to regain a state of operating equilibrium after being subjected to a physical disturbance, with most system variables bounded so that practically the entire system remains intact" [6]. Power system stability phenomena are known as *rotor angle stability*, *voltage stability*, and *frequency stability*.

Rotor angle stability is the ability of the power system to maintain synchronization after being subjected to a disturbance. In case of transient (large disturbance) angle stability, a severe disturbance does allow a generator to deliver its output electricity power into the network. Small-signal (steady-state) angle stability is the ability of the power system to maintain synchronization under small disturbances. The rotor angle stability has been fairly guaranteed by power system stabilizers (PSS), thyristor exciters, fast fault clearing, and other stability controllers and protection actions such as generator tripping.

Voltage stability is the ability of a power system to maintain steady acceptance voltages at all system buses after being subjected to a disturbance from an assumed initial equilibrium point. A system enters a state of voltage instability when a disturbance changes the system condition to make a progressive fall or rise of voltages of some buses. Loss of load in an area, tripping of transmission lines, and other protected equipment are possible results of voltage instability.

Frequency stability is the ability of a power system to maintain system frequency within the specified operating limits. Generally, frequency instability is a result of a significant imbalance between load and generation, and it is associated with poor coordination of control and protection equipment, insufficient generation reserves, and inadequacies in equipment responses.

The size of disturbance, physical nature of instability, the dynamic structure, and the time span are important factors to determine the instability form. The above instability classification is mainly based on dominant initiating phenomena. Each instability form does not always occur in its pure form. One may lead to the other, and the distinction may not be clear. However, distinguishing between different instability forms is important in understanding the underlying causes of the problem in order to develop appropriate design and operating procedures.

To maintain system stability and desirable performance, numerous control loops are in use in a power system. Power system controllers are of many types with different control tasks, including generation excitation controls, prime mover controls, generator/load tripping, fast fault clearing, high speed reclosing, reactive power compensation, load-frequency control, current injection, fast phase angle control, and high-voltage DC (HVDC) special controls.

Power system controls attempt to return the system from an off-normal operating state to a normal operating state. Classifying the power system operating states to *normal*, *alert*, *emergency*, *in extremis*, and *restorative* is conceptually useful for designing appropriate control systems. In the normal state, all system variables (e.g., voltage and frequency) are within the normal range. In the alert state, all system variables are still within the acceptable range. However, the system may be ready to move into the emergency state following a disturbance. In the emergency state, some system variables are outside of the acceptable range and the system is ready to fall into the *in extremis* state. Partial or system wide blackout could occur in the *in extremis* state. Finally, energizing the system or its parts and reconnecting/resynchronizing of system parts occurs during the restorative state.

From the operating state point of view, power system controls can be mainly divided into two different categories: normal/preventive controls, which are applied in the normal and alert states to stay in or return into normal condition, and emergency controls, which are applied in emergency or in *in extremis* state to stop the further progress of the failure and return the system to a normal or alert state.

Automatic frequency and voltage controls are part of the normal and preventive controls, while some control schemes like underfrequency load shedding (UFLS), undervoltage load shedding (UVLS), and special system protection plans can be considered as emergency controls. Control command signals for normal/preventive controls usually include active power generation set points, flow controlling reference points (FACTS), voltage set point of generators, static VAR compensator (SVC), reactor/capacitor switching, and so on. Emergency control measures are some control commands such as tripping of generators, shedding of load blocks, opening of interconnection to neighboring systems, and blocking of transformers tap changer.

Most control loops such as prime mover and excitation controls operate directly on the generation site, and are located at the power plants. In a power plant, the governor voltage and reactive power output are regulated by excitation control, while energy supply system parameters (temperatures, flows, and pressures) and speed regulation are performed by prime mover controls. Automatic generation control balances the total generation and load (plus losses) to reach the nominal system frequency and scheduled power interchange with neighboring systems.

Furthermore, there are many controls and protection systems on the demand side in transmission and distribution networks such as switching capacitor/reactors, tap-changing/phase shifting transformers, HVDC controls, synchronous condensers, and static VAR compensators. Despite numerous existing nested control loops that control different quantities in the system, working in a secure attraction region with a desired performance is the objective of an overall power system control strategy.

For the purpose of dynamic analysis and control synthesis, it is noteworthy to know the timescale of various control loops. The timescale of interest for rotor angle stability in transient (large disturbance) stability studies is usually limited to 3–10 s, and in steady-state (small-signal) studies is on the order of 10–20 s. The rotor angle stability is known as a short-term stability problem, while the voltage stability problem can be either a short-term or a long-term stability problem. The time frame of interest for voltage stability problems may vary from a few seconds to several minutes. Although power system frequency stability is impacted by fast as well as slow dynamics, the time frame will range from a few seconds to several minutes [7]. Therefore, the frequency stability is known as a long-term stability problem.

For the purpose of power system control designs, generally the control loops at lower system levels (locally in a generator) are characterized by smaller time constants than the installed control loops at a higher system level. For example, the automatic

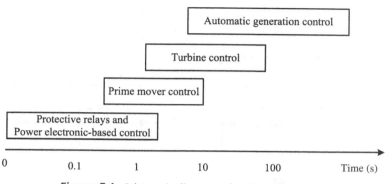

Figure 5.1. Schematic diagram of control timescales.

voltage regulator (AVR), which regulates the voltage of the generator terminals to the reference value, responds typically in a timescale of a second or less. The secondary voltage control, which determines the reference values of the voltage controlling devices, among which are the generators, operates in a timescale of several seconds or minutes. That means these two control loops are virtually decoupled. A schematic diagram showing different control timescales is presented in Fig. 5.1.

5.2 ANGLE AND VOLTAGE CONTROL

As mentioned, angles of nodal voltages (rotor/power angles), nodal voltage magnitudes, and network frequency are three important quantities for power system operation and control. They are also significant in the stability classification point of view. This section is focused on angle and voltage stability, which can be divided into small- and large-disturbance stability. Angle and voltage stability refer to damping of power swings inside subsystems and between subsystems on an interconnected grid and voltage excursion during variation beyond specified threshold levels, respectively.

The risk of losing angle and voltage stability can be significantly reduced by using proper control devices inserted into the power system to find a smooth shape for the system dynamic response. Important control devices for stability enhancement are known as PSS, AVR, and FACTS devices.

The generators are usually operated at constant voltage by using an AVR, which controls the excitation of the machine via the electric field exciter. The exciter supplies the field winding of the synchronous machine with direct current to generate the required flux in the rotor.

A PSS is a controller, besides being the turbine-governing system, performs an additional supplementary control loop to the AVR system of a generating unit. A common structure for the PSS–AVR is shown in Fig. 5.2. There are a number of possible ways for constructing the PSS–AVR system, in which a particular case is already

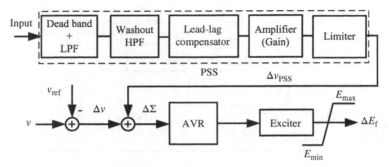

Figure 5.2. PSS and AVR control loops.

introduced in Fig. 4.5. The necessity of the supplementary control loop is due to the conflict behavior of rotor speed and voltage dynamics.

In the steady state, Δv_{PSS} must be equal to zero so that it does not distort the voltage regulation process. However, in the transient state, the generator speed is not constant, the rotor swings, and ΔV undergoes variations caused by the change in rotor angle [8,9]. This voltage variation is compensated by the PSS providing a damping signal Δv_{PSS} that is in phase with generator speed change ($\Delta\omega$).

As shown in the general structure of the PSS (Fig. 5.2), the input signal is passed through a combination of low- and high-pass filters. To provide the required amount of phase shift, the prepared signal is then passed through a lead-lag compensator. Finally, the PSS signal is amplified and limited to provide an effective output signal (Δv_{PSS}). Typically, the rotor speed/frequency deviation ($\Delta\omega/\Delta f$), the generator active power deviation (ΔP_e), or a combination of rotor speed/frequency and active power changes can be considered as input signal to the PSS.

In many power systems, advanced measurement devices and modern communications are already being installed. Using these facilities, as mentioned in Section 4.1, the parameters of the PSS and AVR can be adjusted using an online monitoring-based tuning mechanism. This control architecture is simply shown in Fig. 5.3.

Like frequency control, voltage control is also characterized via several control loops on different system levels. The AVR loop, which regulates the voltage of generator terminals, is located on lower system levels and responds typically in a timescale of a second or less. On the other hand, the secondary voltage control, which determines the voltage reference values of the distributed voltage compensators (e.g., AVR), is activated on a higher system level and operates in a timescale of tens of seconds or minutes. Secondary voltage control is required to coordinate adjustment of the set points of the AVRs and other reactive power sources in a given network to enhance voltage stability of the grid.

The voltage stability can be further enhanced with the use of a higher control level (with a timescale of several minutes) known as tertiary voltage control, based on the overall grid economic optimization. A typical generic of the mentioned three voltage control levels is discussed in Ref. [7].

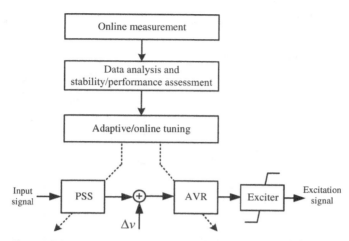

Figure 5.3. A structure for advanced PSS–AVR tuning approaches.

5.3 FREQUENCY CONTROL [10,11]

A severe system stress resulting in an imbalance between generation and load seriously degrades the power system performance (and even stability), which cannot be described in conventional transient stability and voltage stability studies. This type of usually slow phenomena must be considered in relation with the power system frequency control issue.

Frequency deviation is a direct result of an imbalance between the electrical load and the power supplied by the connected generators, so it provides a useful index to indicate the generation and load imbalance. A permanent off-normal frequency deviation may affect power system operation, security, reliability, and efficiency by damaging equipment, degrading load performance, overloading transmission lines, and triggering the protection devices.

Since the frequency generated in an electric network is proportional to the rotation speed of the generator, the problem of frequency control may be directly translated into a speed control problem of the turbine-generator unit. This is initially overcome by adding a governing mechanism that senses the machine speed and adjusts the input valve to change the mechanical power output to track the load change and to restore frequency to the nominal value. Depending on the frequency deviation range, different frequency control loops may be required to maintain power system frequency stability.

The typical frequency control loops are simply represented in Fig. 5.4. Under normal operation, small frequency deviations can be attenuated by the *primary control*. For larger frequency deviation (off-normal operation), according to the available amount of power reserve, the *secondary control*, which is known as load-frequency control (LFC) is responsible for restoring system frequency. The LFC is the main component of automatic generation control (AGC). However, for a serious load–generation imbalance associated with rapid frequency changes following a significant fault, the LFC system may be unable to restore frequency. In this situation, another action must be applied using *tertiary control*,

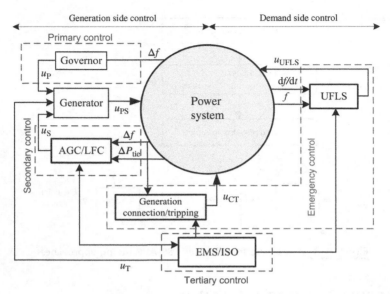

Figure 5.4. Frequency control loops.

standby supplies, or *emergency control* and protection schemes (e.g., UFLS) as the last option to decrease the risk of cascade faults, additional generation events, and load/network and separation events. The tertiary and emergency controls could be realized by the energy management system (EMS), independent system operator (ISO), or market operator.

In a power system, all four forms of frequency control are usually present. The demand side can also participate in frequency control through the action of frequency-sensitive relays, which disconnect some loads at given frequency thresholds (in the UFLS) or using self-regulating effect of frequency-sensitive loads, such as induction motors. Figure 5.5 shows the discussed frequency control loops and corresponding power reserves. The amount of required power reserve depends on several factors including the type and size of load–generation imbalance.

Primary frequency control loop provides a local and automatic frequency control by adjusting the speed governors in the time frame of seconds after a disturbance. The secondary frequency control loop initializes a centralized and automatic control task using the assigned spinning reserve, which is activated in the time frame of a few seconds to minutes after a disturbance. The tertiary frequency control is usually known as a manual frequency control by changing the dispatching of generating units, in the timescale of tens of minutes up to hours after a disturbance.

In the conventional power grids, the primary control reserves a maximum duration of 30 s, whereas in the modern power grids and microgrids with lower inertia, the time constants are much smaller. Virtual inertia can be considered as an effective solution to support the primary frequency control and compensate for the fast frequency changes.

The typical frequency control loops are represented in Fig. 5.4, in a simplified scheme. In a large multiarea power system, all four forms of frequency control (primary,

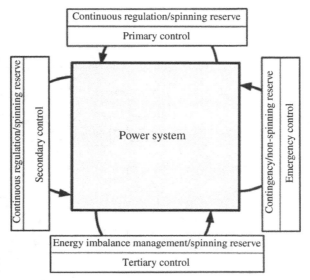

Figure 5.5. Frequency control levels and power reserves.

secondary, tertiary, and emergency) are usually present. The demand side also participates in the frequency control through the action of frequency-sensitive relays, which disconnect some loads at the given frequency thresholds (in the UFLS). The demand side may also contribute to frequency control using the self-regulating effect of frequency-sensitive loads, such as induction motors. However, this type of contribution is not always taken into account in the calculation of the overall frequency control response.

5.3.1 Frequency Control Dynamic

In addition to the primary frequency control, a large synchronous generator may be equipped with the secondary frequency control loop. A frequency response model for control area i in a multiarea power system is shown in Fig. 5.6. In this figure, H_i, D_i, β_i, and Δf_i are the area's constant inertia, damping coefficient, bias factor, and frequency deviation, respectively. $\Delta P_{\text{tie-}i}$ indicates the tie-line power change of area i, and is the synchronizing torque coefficient between area i and area j. Here, $M_{ki}(s)$, R_{ki}, and α_{ki} are the governor-turbine model, droop characteristic, and participation factor for generator unit k in area i, respectively.

The synchronous generators are equipped with primary and secondary frequency control loops. The secondary loop performs a feedback via the frequency deviation and adds it to the primary control loop through a dynamic *controller*. The resulting signal (ΔP_C) is used to regulate the system frequency. In real-world power systems, usually the dynamic controller is a simple integral (I) or proportional-integral (PI) controller. Following a change in the load, the feedback mechanism provides an appropriate signal for the turbine to make generation (ΔP_m) track the load and restore the system frequency.

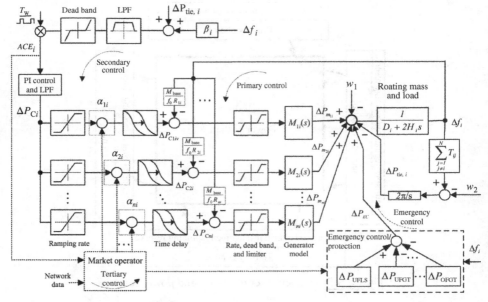

Figure 5.6. Frequency response model.

In addition to the area frequency regulation, the secondary control loop should control the net interchange power with neighboring areas at scheduled values. This is generally accomplished by feeding a linear combination of tie-line flow and frequency deviations, known as *area control error* (ACE), via secondary feedback to the dynamic controller. The ACE can be calculated as follows:

$$\text{ACE}_i = \Delta P_{\text{tie},i} + \beta_i \Delta f_i \tag{5.1}$$

where bias factor β_i can be computed as

$$\beta_i = \frac{1}{R_i} + D_i \tag{5.2}$$

The block diagram shown in Fig. 5.6 illustrates how Equation (5.1) is implemented in the secondary frequency control loop. The effects of local load changes and interface with other areas are also considered as the following two input signals:

$$w_1 = \Delta P_{Li}, w_2 = \sum_{\substack{j=1 \\ j \neq i}}^{N} T_{ij} \Delta f_j \tag{5.3}$$

where ΔP_{Li} and T_{ij} are the load disturbance (in area i) and synchronizing coefficient of two areas (i and j), respectively.

As mentioned, the secondary loop performance is highly dependent on how the participant generating units would respond to the control action signals. The North

American Electric Reliability Council separated generator actions into two groups. The first group is associated with large frequency deviations where generators would respond through governor action and the second group is associated with a continuous regulation process in response to the secondary frequency control signals only. During a sudden increase in area load, the area frequency experiences a transient drop. At the transient state, there would be flows of power from other areas to supply the excess load in that area. Usually, certain generating units within each area would be on regulation to meet this load change. At steady state, the generation would be closely matched with the load, causing tie-line power and frequency deviations to drop to zero.

In an interconnected power system the control area concept needs to be used for the sake of synthesis and analysis of the secondary frequency control system. A control area is an electric network that is managed under a common automatic control scheme maintaining frequency and tie-line interchanges close to the specified nominal values by balancing load and generation. The frequency is assumed to be the same in all points of a control area.

In practice, to clear the fast changes and probable added noises, system frequency gradient and ACE signals must be filtered before being used in the secondary frequency control loop. If the ACE signal exceeds a threshold at interval T_W, it will be applied to the controller block. The controller can be activated to send higher/lower pulses to the participant generation units if its input ACE signal exceeds a standard limit. Delays, ramping rate, and range limits are different for various generation units. Concerning the limit on generation, governor dead band and time delays, the LFC model becomes highly nonlinear; hence, it will be difficult to use the conventional linear techniques for performance optimization and control design [5].

For the purpose of frequency control synthesis and analysis in the presence of load disturbances, a simple low-order linearized model is commonly used. The overall generation–load dynamic relationship between the incremental mismatch power $(\Delta P_m - \Delta P_L)$ and the frequency deviation (Δf) can be expressed as

$$\Delta P_m(t) - \Delta P_L(t) = 2H \frac{d\Delta f(t)}{dt} + D\Delta f(t) \tag{5.4}$$

where ΔP_m is the mechanical power change, ΔP_L is the load change, H is the inertia constant, and D is the load damping coefficient. Using the Laplace transform, Equation (5.4) can be written as

$$\Delta P_m(s) - \Delta P_L(s) = 2Hs\Delta f(s) + D\Delta f(s) \tag{5.5}$$

Equation (5.5) is represented on the right-hand side of the frequency response model in Fig. 5.6.

In a multiarea power system, the trend of frequency measured in each control area is an indicator of the trend of the mismatch power in the interconnection and not in the control area alone. Therefore, the power interchange should be properly considered in the frequency response model. It is easy to show that in an interconnected power system with N control areas, the tie-line power change between area i and other areas can be represented as [5]

$$\Delta P_{\text{tie},i} = \sum_{\substack{j=1 \\ j \neq i}}^{N} \Delta P_{\text{tie},ij} = \frac{2\pi}{s} \left[\sum_{\substack{j=1 \\ j \neq i}}^{N} T_{ij}\Delta f_i - \sum_{\substack{j=1 \\ j \neq i}}^{N} T_{ij}\Delta f_j \right] \qquad (5.6)$$

Equation (5.6) is realized on the right-hand side of the frequency response block diagram in Fig. 5.6. The effect of changing the tie-line power for an area is equivalent to changing the load of that area. That is why, $\Delta P_{\text{tie},i}$ has been added to the mechanical power change (ΔP_{m}) and area load change (ΔP_{L}) using an appropriate sign in Fig. 5.6.

Each control area monitors its own tie-line power flow and frequency at the area control center, and the combined signal (ACE) is allocated to the dynamic controller. Finally, the resulting control action signal is applied to the turbine-governor units, according their participation factors.

The *participation factor* indicates the amount of participation of a generator unit in the secondary frequency control system. Following a load disturbance within the control area, the produced appropriate control signal is distributed among generator units in proportion to their participation, to make generation follow the load. In a given control area, the sum of participation factors is equal to 1.

$$\sum_{k=1}^{n} \alpha_{ki} = 1, \qquad 0 \leq \alpha_{ki} \leq 1 \qquad (5.7)$$

In a competitive environment, the participation factors are actually time-dependent variables and must be computed dynamically based on bid prices, availability, congestion problems, costs, and other related issues [5,10].

As mentioned, in the case of a large generation loss disturbance, the scheduled power reserve may not be enough to restore the system frequency, and the power system operators may follow an emergency control plan such as UFLS. The UFLS strategy is designed so as to rapidly balance the demand of electricity with the supply and to avoid a rapidly cascading power system failure. Allowing normal frequency variations within expanded limits will require the coordination of primary control and scheduled reserves with generator load set points; for example, underfrequency generation trip (UFGT), overfrequency generation trip (OFGT), or overfrequency generator shedding (OFGS) and other frequency-controlled protection devices.

In the case of contingency analysis, the emergency protection and control dynamics must be adequately modeled in the frequency response model. Since they influence the power generation–load balance, the mentioned emergency control dynamics can be directly included to the system frequency response model. This is made by adding an emergency protection/control loop to the primary and secondary frequency control loops, as shown in Fig. 5.6. The $\Delta P_{\text{UFLS}}(s)$, $P_{\text{UFGT}}(s)$, and $\Delta P_{\text{OFGT}}(s)$ represent the dynamics effects of the UFLS, UFGT, and OFGT actions, respectively.

The emergency control schemes and protection devices dynamics are usually represented using incremented/decremented step behavior. Thus, in Fig. 5.6, for simplicity, the related blocks can be represented as a sum of incremental (decremental) step functions. For instance, for a fixed UFLS scheme, the function of ΔP_{UFLS} in the time domain could be considered as a sum of the incremental step functions of $\Delta P_j u(t - t_j)$. Therefore, for L load shedding steps

$$\Delta P_{\text{UFLS}}(t) = \sum_{j=0}^{L} \Delta P_j u(t - t_j) \tag{5.8}$$

where ΔP_j and t_j denote the incremental amount of load shed and time instant of the jth load shedding step, respectively. Similarly, to formulate the ΔP_{OFGT}, ΔP_{UFGT}, and other emergency control schemes, appropriate step functions can be used. Therefore, using the Laplace transformation, it is possible to represent $\Delta P_{\text{EC}}(s)$ in the following summarized form:

$$\Delta P_{\text{EC}}(s) = \sum_{l=0}^{N} \frac{\Delta P_l}{s} e^{-t_l s} \tag{5.9}$$

where ΔP_l is the size of equivalent step load/power changes due to a generation–load event or a load shedding scheme at t_l.

The effect of tertiary control loop in a control unit (e.g., supervisory control and data acquisition/dispatching center) organized by the market operator in relation with Gencos and security plans is also shown in Fig. 5.6.

5.3.2 Operating States and Power Reserves

Frequency provides a useful index to indicate the system generation and load imbalance. Any short-term energy imbalance results in an instantaneous change in system frequency as the disturbance is initially offset by the kinetic energy of a rotating plant. As mentioned, off-normal frequency deviations can impact the power system operation, system reliability, and efficiency. Large frequency deviations can damage equipment, degrade load performance, overload transmission lines, and trigger the protection relays and may ultimately lead to a system collapse.

Depending on the size of the frequency deviation experienced, primary, secondary, tertiary, and emergency controls may all be required to maintain power system frequency. One method of characterizing frequency deviations experienced by a power system is in terms of frequency deviation ranges and related control actions as shown in Table 5.1.

The frequency variation ranges Δf_1, Δf_2, Δf_3, and Δf_4 are identified in terms of different power system operating conditions, specified in terms of local regulations. Under normal operation, frequency is maintained near the nominal frequency by balancing generation and load. That is, the small frequency deviations up to Δf_1 can be attenuated by the governor natural autonomous response (primary control). The secondary control can be used to restore area frequency deviation of more than Δf_1. In particular, a secondary control system must be designed to maintain the system frequency and tie-line power deviations within the limits of specified operating

TABLE 5.1. Frequency Deviation Ranges and Associated Control Actions

Δf	Condition	Control Action	Reserve
Δf_1	No contingency or load event	Normal operating	Continuous/spinning
Δf_2	Generation–load or network event	AGC operating	Continuous/spinning
Δf_3	Contingency/separation event	Tertiary/emergency operating	Contingency/spinning, Nonspinning
Δf_4	Multiple contingency event	Emergency operating	Nonspinning/load shedding

standards, using the available spinning reserves. The value of Δf_2 is mainly determined by the available amount of operating reserved power in the system.

For large imbalances in real power associated with large/rapid frequency changes (e.g., Δf_3 and Δf_4 frequency deviation events) following a severe disturbance/fault, the secondary control loop may be unable to recover the system frequency. In this case, to prevent additional generation events, load/network events, separation events or multiple contingency events, the tertiary and/or emergency controls, as well as other protection schemes must be used to restore the system frequency.

According to the UCTE commitment [12], a sudden loss of 3000 MW of generating capacity must be offset by primary control alone, without the need for load shedding. Likewise, the sudden load shedding of 3000 MW in total must not lead to a frequency deviation exceeding Δf_1 (180 mHz).

In order to have a reliable and secure operation, enough regulation power reserve should be available, so that the ACE, the instantaneous difference between the actual and the reference values for the power interchange of a control area, is kept within reasonable bounds. Power reserves ensure that capacity is available when needed to maintain secure power system operations following an imbalance in the load–generation system.

In a power system market, the reserves must be carefully planned and purchased so that the system operator is able to use them when required. The system operator guarantees effective use of assets, including the dispatch of energy and the dispatch of spinning (regulation) and nonspinning (contingency) reserves, and organizes the energy and ancillary services markets. The market operator must activate these power reserves to meet the standard performance indices in a timely and economical manner.

The reserves required during normal conditions are known as spinning reserve or regulation reserve, and are used for continuous regulation and energy imbalance management. This reserve is used to track minute-to-minute fluctuations in the system load–generation pattern, and is provided by online resources with automatic controls that respond rapidly to the raise/down control command signal.

The spinning reserve can be simply defined as the difference between capacity and existing generation level. It refers to spare power capacity to provide the necessary regulation power for the sum of primary and secondary control issues. The time response of spinning reserve for primary frequency regulation is about 30 s, which is much faster than the spinning reserve time response for secondary frequency control (within 15 min). Regulation power is the required power to bring the system frequency back to its nominal value. The frequency-dependent reserves are automatically activated by the AGC system, when the frequency is in a lower level than the nominal value (50 or 60 Hz depending on the system).

Another type of spinning reserve is known as the energy imbalance management reserve that serves as a bridge between the regulation service and the hourly or half-hourly bid-in energy schedules, similar to but slower than continuous regulation. It is used for load-following problems in the tertiary level of frequency control issue. It also serves a financial/settlement function in clearing spot markets [12]. The energy imbalance management reserve must be available within 30 min at a specified minimum rate, typically 2 MW/min.

The nonspinning reserves are instantaneous contingency or replacement reserves that are used during system contingencies. A contingency is a trip of a transmission line or generator, a loss of load, or some combination of these events. This contingency in turn causes other problems, such as a transmission line overload, and significant frequency/voltage deviations or frequency/voltage instability. Contingency reserves are a special percentage of generation capacity resources held back or reserved to meet emergency needs. The contingency reserve services are often referred to as operating reserves. Concerning the frequency control issue, the nonspinning reserves can be classified into two categories [12]: instantaneous contingency reserves and replacement reserves.

Instantaneous contingency reserves (also referred to as the contingency nonspinning reserves) are provided by online generating units (e.g., pumped storage power stations) that are able to rapidly increase output or decrease consumption after receiving a control command in response to a major disturbance or other contingency event. The time response of this reserve, which is known as a quick-start operating reserve, is within 10–15 min.

The replacement reserves will be provided by generating units (e.g., combined-cycle gas-turbine power plants) with a slower response time (up to 30 min) that can be called upon to replace or supplement the instantaneous contingency reserve as standby reserves following receiving a dispatching command or control action signal from the tertiary or emergency control levels. These reserves are typically activated in the case of a generator outage or power imbalances caused by severe events. The instantaneous contingency and replacement reserves are usually activated by the central system operator through a manual control process, while the spinning reserves are usually activated automatically.

Figure 5.5 shows the discussed frequency control levels and corresponding power reserves. In order to determine the sufficient amount of power reserves for proper load–generation balance control with acceptable reliability, it is necessary to refer to the existing reliability standards and assigned performance indices by the relevant technical committees. The amount of required power reserve depends on several factors, including the type and size of imbalance (load–generation variation).

5.4 SUPERVISORY CONTROL AND DATA ACQUISITION

Considering the change of the power market from centralized to decentralized decision making as well as separation of the power market from power system reliability, control centers have to be modified to cope with the changing of the power system environment.

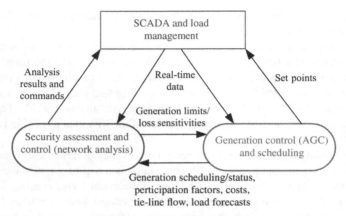

Figure 5.7. Application layer of a modern EMS.

In a modern power system, the supervisory control and data acquisition (SCADA) has an important role in successful operation and control, particularly in the energy management system (EMS). The SCADA together with security control, AGC, and load management are the major units in the application layer of a modern EMS [13]. The AGC process is performed in a control center remote from generating plants, while power production is controlled by turbine governors at the generation site. The AGC communicates with SCADA, load management unit, and security assessment and control center in the EMS, as shown in Fig. 5.7 [14].

The SCADA block covers a number of applications including analog, status and accumulator data processing, limit checking, data processing, historical data recording, tagging, control actions, and load shedding. The generation control and scheduling block consist of reserves monitoring, AGC performance evaluation, transaction scheduling, market interface, and load forecasting. Finally, the security assessment and control system includes topology processing, state estimation, real-time stability assessment, loss sensitivities, contingency analysis, security enhancement, optimal power flow calculation, off-line stability evaluation, and disturbance/fault analysis.

Security assessment and control system includes methods to evaluate power system stability in different conditions and provide the necessary control actions. The main components of this system according to CIGRE Report No. 325 are shown in Fig. 5.8. Measurements of power system quantities and devices status are collected by SCADA

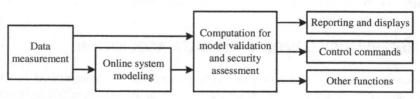

Figure 5.8. Structure of security assessment and control system.

and individual PMU/intelligent electronic devices (IEDs) in the data measurement block. The collected data are used to identify the power system parameters in the online system modeling block. Then, the computation block determines the validation/accuracy of the obtained models as well as the overall security assessment. Finally, the results are used for reporting, monitoring, preventive/corrective controls, as well as other functions such as archiving and modes studying.

A simplified architecture for the SCADA/EMS center and other important connected blocks is shown in Fig. 5.9. System data are collected from remote terminal units (RTUs), IEDs, and integrated substation automation systems (SASs) by using standard communication protocols (e.g., IEC-60870-5-101). The information is exchanged over the Internet and/or via interutility control center communication protocol (ICCP) such as 60870-6-TASE2. The SCADA/EMS also includes a fault/disturbance recording system, a historical information system, and several monitoring and analysis tools that enable the operators to view and organize the system operation.

The SCADA system consists of a master station to communicate with the RTUs and IEDs for a wide range of monitoring and control processes across a power system. In a modern SCADA system, the monitoring, processing, and control functions are distributed among various servers and computers that communicate in the control center using a real-time local area network (LAN). A simplified SCADA center is shown in Fig. 5.10. Although nowadays many data processing and control functions are moved to the IEDs, the power systems still need a master station or control center to organize/coordinate various applications.

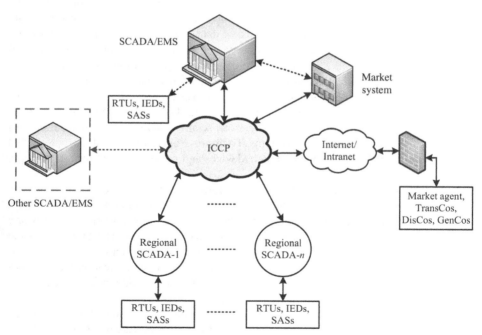

Figure 5.9. Conceptual overview of SCADA/EMS structure.

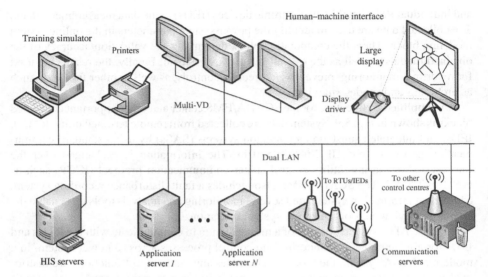

Figure 5.10. A simplified structure for a typical SCADA center.

As shown in Fig. 5.10, the human–machine interface (HMI), application servers, and communication servers are the major elements of the SCADA system. The HMI consists of a multivideo display (Multi-VD) interface and a large display or map-board/ mimic-board to display an overview of the power system. The application servers are used for general database, historical database, data processing, real-time control functions, EMS configuration, and system maintenance. The communication servers are used for data acquisition from RTUs/IEDs, and data exchange with other control centers.

The data communication, system monitoring, alarms detection, and control commands transmission are the common actions in a SCADA center. Moreover, the SCADA/ EMS system performs load shedding and special control schemes in cooperation with the AGC system and security control unit. Various security methods and physical options can be applied to protect SCADA systems. To improve operation security, usually a dual configuration for the operating computers/devices and networks in the form of primary and standby configuration is used.

In a modern SCADA/EMS station, the performed control and monitoring processes are highly distributed among several servers, monitors, and communication devices. Using a distributed structure has many advantages such as easily upgrading of hardware/ software parts, reducing costs, and limiting the effect of failures. The SCADA system uses an open architecture for communication with other systems, and to support interfaces with various vendors' products [15]. A mix of communication technologies such as wireless, fiber optics, and power line communications could be a viable solution in a SCADA system.

As mentioned, modern communications are already being installed in many power systems. Substations at both transmission and distribution levels are being equipped with advanced measurement and protection devices as well as new SCADA systems for

Figure 5.11. A regional SCADA, West Regional Electric Co., Kermanshah.

supervision and control. Communication between control units is also being modernized as is the communication between several subsystems of the high level control at large power producers at the EMS level. These are often based on open protocols, notably the IEC61850 family for SCADA-level communication with substations and distributed generating units, and the IEC61968/61970 CIM family for EMS-level communication between control centers [16].

In some cases, the role of a SCADA system is distributed between several regional SCADAs; usually, one of them is a coordinator and works as the master SCADA center. A real view of a regional SCADA is shown in Fig. 5.11.

In real-power system structures, the SCADA/EMS effectively uses IEDs for doing remote monitoring and control actions. The IEDs as monitoring and control interface to the power system equipment can be installed in remote (site/substation) control centers and can be integrated using suitable communication networks. This issue facilitates to accomplish a remote site control system similar to the major station in the SCADA/EMS. A simplified architecture is presented in Fig. 5.12.

A remote site control center may consist of an RTU, IEDs, HMI database server, and synchronizing time generator. The RTU and IEDs are for communication with the SCADA/EMS station, remote access control functions, data measurement/concentration, and status monitoring. The synchronizing time generator is typically a GPS satellite clock that distributes a time signal to the IEDs.

The local access to the IEDs and the local communication can be accomplished over a LAN, while, the remote site control center is connected to the SCADA/EMS, EMS, and other engineering systems through the power system wide area network. The interested readers can find relevant standards for SCADA/EMS systems, substation automation, remote site controls with detailed architectures, and functions of various servers, networks, and communication devices in Ref. [15].

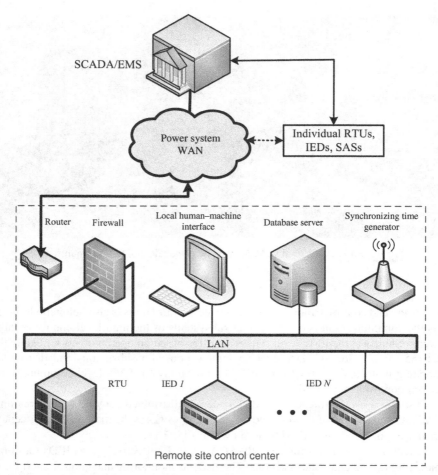

Figure 5.12. A simplified architecture including remote site control and SCADA system.

5.5 CHALLENGES, OPPORTUNITIES, AND NEW PERSPECTIVES

5.5.1 Application of Advanced Control Methods and Technologies

In the last 20 years, intelligent systems applications have received increasing attention in various areas of power systems such as operation, planning, control, and management. Numerous research papers indicate the applicability of intelligent techniques to power systems. While many of these systems are still under investigation, there already exist a number of practical implementations of intelligent systems in many countries across the world. In conventional schemes, power system operation, planning, control, and management are based on human experience and mathematical models to find solutions; however, power systems have many uncertainties in practice. Namely, those mathematical models provide only for specific situations of the power systems under respective assumptions. With these assumptions, the solutions of power system control analysis/synthesis problems are not trivial. Therefore, there exist some limitations for the mathematical model based schemes. In order to overcome these limitations, applications of intelligent technologies such as knowledge-based expert systems, fuzzy systems, artificial neural networks, genetic algorithms, Tabu search, and other intelligent technologies have been investigated in a wide area of power system problems to provide a reliable and high-quality power supply at minimum cost. In addition, recent research works indicate that more emphasis has been put on the combined usage of intelligent technologies for further improvement of the operation, control, and management of power systems.

Several surveys on the worldwide application of intelligent methodologies on power systems have been recently published. Intelligent and advanced systems are currently used in many utilities across the developed countries. Some application areas of intelligent technologies in Japanese power system utilities are presented in Ref. [10]. Many applications have been proposed in literatures in those areas to demonstrate the advantages of intelligent systems over conventional systems. A certain number of actual implementation of intelligent systems is already working toward better and more reliable solutions for control and operation problems in power systems.

As mentioned earlier, there already exists a quite good number of implementations of intelligent systems in Japanese utilities. Some of them are now at their renewal stages. However, because of the reasons listed in Table 5.2, the renewals of some of the intelligent systems will be postponed. Most of these obstacles will be solved by further developments of software/hardware technologies.

Currently, power system operation and control in all aspects is undergoing fundamental changes due to restructuring, expanding of functionality, rapidly increasing amount of renewable energy sources, and emerging of new types of power generation and consumption technologies. This issue opens the way to realizing new/powerful intelligent techniques.

TABLE 5.2. Problems for Future Extension of Intelligent Systems in Real Power Systems

Amount of additional investment
Cost of maintenance
Unsatisfactory performance
Required processing speed
Shortage of actual operation
Black box based operation
Accuracy of solutions
Acceptability by human operator
Industry owners are too conservative!

5.5.2 Standards Updating

Power system operation is always in a changing state due to the integration of new power sources, maintenance schedules, unexpected outages, and changing interconnection schedules and fluctuations in demand, generation, and power flow over transmission lines. Increasing size, restructuring, emerging of renewable energy sources (RESs), and new uncertainties provides a variable nature for power systems, and it imposes the necessity for continuously updating operating standards. In this direction, frequency and voltage performance standards compliance verification remains a major open issue. Interconnection procedures and standards should be also reviewed to ensure that operating control schemes and their responses are in a consistent manner in all power generation technologies, specifically RESs and variable generation technologies. This is because uncertainty and variability are their two major attributes that distinguish them from conventional forms of generation and may impact the overall system planning and operations.

Therefore, the control issues and related standards may evolve into new guidelines. The standards redesign must be done in both normal and abnormal conditions, and they should allow for the introduction of renewable power generation and modern distributed generator technologies.

5.5.3 Impacts of Renewable Energy Options

There is a rising interest on the impacts of RESs on power system operation and control, as the use of RESs increases worldwide. The RESs certainly affect the dynamic behavior of the power system in a way that might be different from conventional generators. High renewable energy penetration in power systems may increase uncertainties during abnormal operation, and introduces several technical implications and opens important questions as to what happens to each control requirement in case of adding numerous RESs to the existing generation portfolio, and to whether the traditional power system control approaches to operation in the new environment are still adequate.

When renewable power plants are introduced into the power system, an additional source of variation is added to the already variable nature of the system. To analyze the variations caused by RES units, the total effect is important, and every change in RES power output does not need to be matched one by one via a change in another generating unit moving in the opposite direction. However, as already mentioned, the slow RES power fluctuation dynamics and total average power variation may negatively contribute to the power imbalance, which should be taken into account in the new control schemes. Among all RESs, because of dynamic behavior and amount of penetration, the impact of wind power on the system performance is more significant than other types of renewable sources. Therefore, some control schemes may need a revision in the presence of high penetration wind turbines.

It is shown that the doubly fed induction generators (DFIGs) have larger loadability than the induction generators (IGs). Both stator and rotor windings of the IG type wind turbine generators are connected directly with the power grid, but in the DFIG type only stator is directly connected and the rotor is connected through a power electronic type

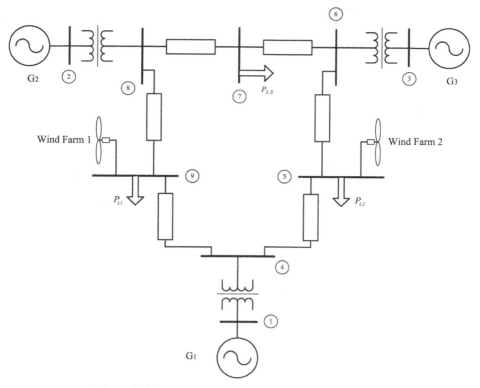

Figure 5.13. Nine-bus test system with two wind farms.

converter. The IG type wind turbine generators in turn add more inertia than DFIG in the power system; and in conclusion, the IG type wind turbine generators' frequency response is better than systems with DFIG type, in the same conditions.

To study the impacts of different types of wind turbines on the voltage and frequency behavior, the IEEE nine-bus power system is considered as a test system [17]. A single diagram of the test system with wind farms is shown in Fig. 5.13. Simulation data and system parameters are given in Appendix B.

As a serious disturbance scenario, the largest generator (G2) in the test system is tripped at $t = 10$ s in the presence of the following conditions: without wind turbine, with 10% DFIG type penetration, with 10% IG type penetration, and with 10% IG type wind turbine compensated with a static compensator (STATCOM). Figure 5.14 shows the frequency and voltage responses following the mentioned disturbance. The rate of frequency change is also illustrated in this figure.

All of the four cases are unstable, but they show different frequency/voltage response behaviors. Here, to protect the system against blackout, an emergency control action may need to be applied at the first few seconds (following the disturbance). The difference between the four test scenarios is clearly demonstrated through the simulation

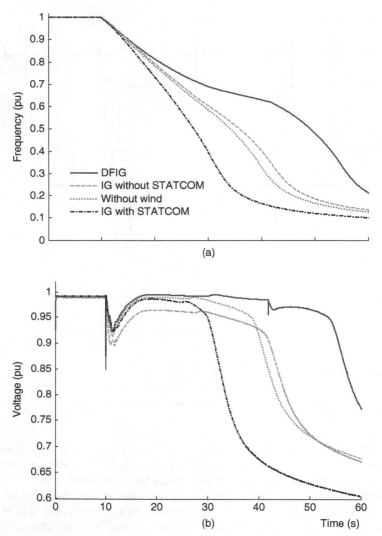

Figure 5.14. System response following G2 outage with and without WTGs: (a) frequency deviation and (b) voltages changes.

results. It is also shown that the frequency and voltage responses may behave in opposite directions.

5.5.4 RESs Contribution to Regulation Services

Planning the required power reserve, concerning the rapid growth of variable renewable generation and the resulting impacts on power system performance also is an important issue in future power system operation and control. For example, consider a power system with a high penetration of wind power. A greater amount of power reserve is needed for a larger amount of fluctuating wind power to cover periods when there is no wind. On the other hand, managing surplus electricity during periods of strong wind could be also considered as a challenge. Contribution of a renewable power plant (e.g., wind farm) in the ancillary/regulation services to provide the regulation power reserve can be considered as a proper solution. In combination with advanced forecasting techniques, it is now possible to design variable generators with the full range of performance capability that is comparable, and in some cases superior, to conventional synchronous generators. For example, unlike a typical thermal power plant whose output ramps downward rather slowly, wind farms can react quickly to a dispatch instruction taking seconds, rather than minutes.

Therefore, variable generation resources, such as wind power facilities, can be equipped to provide governing and participate in regulation tasks as well as conventional generators. For example, in the near future, the RESs are needed to actively participate in frequency control issue and maintaining system reliability along with conventional generation. Figure 5.15 shows the frequency response model for such cooperation [11].

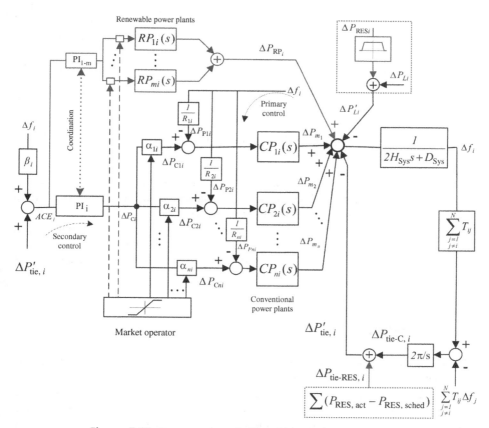

Figure 5.15. Frequency regulation with contribution of RESs.

Here, renewable power plants $(\text{RP}_{1i}(s), \ldots, \text{RP}_{mi}(s))$, such as wind farms that can provide a considerable amount of power, also participate in the frequency regulation system by producing regulation renewable power $(\Delta P_{\text{RP}i})$.

Figure 5.15 shows a block diagram of a typical control area with n conventional and m renewable generating units. Here, Δf is the frequency deviation, ΔP_{m} is the mechanical power, ΔP_{C} is the secondary frequency control action, ΔP_{L} is the load disturbance, H_{Sys} is the equivalent inertia constant, D_{Sys} is the equivalent damping coefficient, β is the control frequency bias, R_i is the drooping characteristic, ΔP_{P} is the primary frequency control action, α_i is the participation factor, ΔP_{RES} is the RES power fluctuation, ACE is the area control error, $M_i(s)$ is the conventional governor-turbine model, and finally, $\Delta P'_{\text{L}}$ and $\Delta P'_{\text{tie}}$ are the augmented local load change and tie-line power fluctuation signals, respectively. Here, for simplicity, the corresponding blocks for GRC, governor dead bands, and time delays (which are shown in Fig. 5.6) as well as the detailed dynamics of renewable power plants are not indicated in Fig. 5.15.

Similar to conventional generating units, the participation factor for each power plant is determined by market operator. These power plants may use individual controllers, but those controllers should be coordinated together, as well as conventional ones. In new frequency response models, the ACE signal should represent the impacts of renewable power on the scheduled flow over the tie-line, as well as the local power fluctuation via the area frequency. According to Fig. 5.15, the updated ACE signal can be obtained as follows [11]:

$$\begin{aligned}
\text{ACE} &= \beta\Delta f + \Delta P'_{\text{tie}} \\
&= \beta\Delta f + \left(\sum(P_{\text{tie-C,act}} - P_{\text{tie-C,sched}}) + \sum(P_{\text{tie-RES,act}} - P_{\text{tie-RES,sched}})\right)
\end{aligned} \tag{5.10}$$

where $P_{\text{tie-C,act}}$, $P_{\text{tie-C,sched}}$, $P_{\text{tie-RES,act}}$, and $P_{\text{tie-RES,sched}}$ are actual conventional tie-line power, scheduled conventional tie-line power, actual RES tie-line power, and scheduled RES tie-line power, respectively. In addition to the wind turbines, other participants can also provide regulation services, such as storage devices that smooth either consumption or generation, consumers that can modulate their consumption upon request or automatically, and to some extent RESs. The demand for AGC services is defined by the market operator and depends on the power system structure.

5.6 SUMMARY

This chapter provides an introduction on the general aspects of power system stability and control. Fundamental concepts and definitions of angle, voltage and frequency stability, and existing controls are emphasized. The timescales and characteristics of various power system controls are described. The SCADA/EMS architecture in modern power grids is explained. Finally, some challenges and new research directions are presented.

REFERENCES

1. C. P. Steinmetz, Power control and stability of electric generating stations, *AIEE Trans.*, **XXXIX**(Part II), 1215–1287, 1920.

2. AIEE Subcommittee on Interconnections and Stability Factors, First report of power system stability, *AIEE Trans.*, 51–80, 1926.

3. EPRI Report EL-6627, Long-term dynamics simulation: modeling requirements. Final Report of Project 2473-22, prepared by Ontario Hydro, 1989.

4. IEEE Special Stability Controls Working Group, Annotated bibliography on power system stability controls: 1986–1994, *IEEE Trans. Power Syst.*, **11**(2), 794–800, 1996.

5. H. Bevrani, *Robust Power System Frequency Control*, Springer, New York, 2009.

6. P. Kundur, J. Paserba, V. Ajjarapu, et al., Definition and classification of power system stability, *IEEE Trans. Power Syst.*, **19**(2), 1387–1401, 2004.

7. G. Andersson, et al., Frequency and voltage control. In: A. Gomez-Exposito, et al., editors, *Electric Energy Systems: Analysis and Operation*, CRC Press, 2009.

8. J. Machowski, et al., *Power System Dynamics: Stability and Control*, 2nd edn, Wiley, Chichester, UK, 2008.

9. P. Kundur, *Power System Stability and Control*, McGraw-Hill, New York, 1994.

10. H. Bevrani and T. Hiyama, *Intelligent Automatic Generation Control*, CRC Press, New York, 2011.

11. H. Bevrani, Automatic generation control. In: H. Wayne Beaty, editor, *Standard Handbook for Electrical Engineers*, 16th edn, McGraw-Hill, New York, 2013, Section 16.8, pp. 138–159.

12. P. Horacek, Securing electrical power system operation. In: S. Y. Nof, editor, *Springer Handbook of Automation*, Springer, 2009, pp. 1139–1163.

13. N. K. Stanton, J. C. Giri, and A. Bose, Energy management. In: L. L. Grigsby, editor, *Power System Stability and Control*, CRC Press, Boca Raton, FL, 2007.

14. S. C. Savulescu, editor, *Real-Time Stability Assessment in Modern Power System Control Centers*, Wiley, New York, 2009.

15. IEEE Standard C37.1, *Standard for SCADA and Automation Systems*, 2008.

16. H. Bindner and O. Gehrke, System control and communication. In: H. Larsen and L. S. Petersen, editors, *Risø Energy Report: The Intelligent Energy System Infrastructure for the Future*, Vol. 8, National Laboratory for Sustainable Energy, Roskilde, Denmark, 2009, pp. 39–42.

17. H. Bevrani and A. G. Tikdari, An ANN-based power system emergency control scheme in the presence of high wind power penetration. In: L. F. Wang, et al., editors, *Wind Power Systems: Applications of Computational Intelligence*, Springer Book Series on Green Energy and Technology, Springer, Heidelberg, 2010.

6

WIDE-AREA MEASUREMENT-BASED POWER SYSTEM CONTROL DESIGN

Interconnections in a power system are intended to improve the system's reliability and economic efficiency. On the other hand, these interconnections occasionally cause interarea low-frequency oscillation with poor damping characteristics as described in Chapter 2. Therefore, many kinds of methods for designing a damping controller, such as a power system stabilizer (PSS), have been developed to damp interarea oscillations as well as local oscillations [1–3].

Recently, real-time monitoring of power systems based on the wide-area phasor measurement [4] has attracted the attention of power system engineers for the state estimation and system protection especially under a deregulated environment with complex power contracts. A proper grasp of the present state with flexible wide-area control should become a key issue in keeping power system stability properly. As mentioned in Chapter 2, the dynamic characteristics of oscillation modes, especially the interarea low-frequency mode with poor damping, can be detected successfully from phasor measurements [5]. Therefore, detected unstable modes can be damped adaptively and effectively by controlling the power system, for example, by tuning controller parameters based on the measurement data as described in Sections 4.1 and 5.2.

Power System Monitoring and Control, First Edition. Hassan Bevrani, Masayuki Watanabe, and Yasunori Mitani.

This chapter is mainly focused on the tuning of PSSs using wide-area phasor measurement. The identified low-order model based on the measurement represents power system oscillations, which can be used to design PSSs and to assess the effectiveness of PSS tuning in the real large-scale power system. Small-signal phasor fluctuations are available for the identification of the oscillation characteristics and the evaluation process for designed controllers. Therefore, damping controllers can be designed adaptively for a specific time interval. As a result, dominant swing modes are effectively damped by tuned controllers suitable for the present power system state. Some numerical analyses demonstrate the effectiveness of the method by using phasor dynamical data obtained by a power system simulation package.

6.1 MEASUREMENT-BASED CONTROLLER DESIGN

Many approaches for the measurement-based controller design have been investigated in the literatures such as the PSS design based on the Prony method [6], the identification of low-order state-space power system models by using multi-input multi-output procedures [7], and the low-order identification technique coupled with a robust controller design technique based on genetic algorithm [8]. These methods require some perturbation input to the system to identify the system state.

Figure 6.1 shows the schematic diagram of the concept of damping controller design based on the wide-area phasor measurements, which is described in this chapter. This diagram can be considered as an example realization scheme for the general real-time measurement-based approach described in Figs. 4.2 and 5.3. The advantage of the tuning scheme is that steady state phasor fluctuations are available for the identification process and the effect assessment of the tuned control. In other words, a large disturbance like a line fault is not necessary since the stability of dominant swing modes can be investigated directly by using eigenvalues of the identified model. Also, the identification process does not require information on the system input for perturbation, while the mentioned system identification methods require both input and output signals of the system.

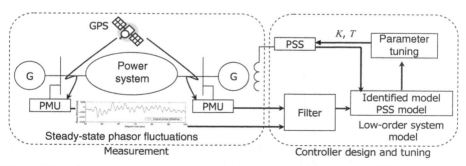

Figure 6.1. Schematic diagram of damping controller design with wide-area measurements.

6.2 CONTROLLER TUNING USING A VIBRATION MODEL

6.2.1 A Vibration Model Including the Effect of Damping Controllers

The power swing equations of generators in an n-machine system can be represented by (3.14). In a multimachine power system multiple swing modes exist and interact with each other. For example, in a longitudinally interconnected power system like the western Japan 60 Hz system, two interarea modes associated with power oscillations tend to be dominant and interact with each other [9]. One of the modes is associated with the oscillation between both end generators; the other is associated with the oscillation among three generators, which are both ends and the middle one. These oscillation characteristics can be explained by right eigenvectors (2.4) and the participation factor (2.6). Note that the following method is applicable to any other system such as a longitudinal system in cases where characteristic oscillations are observed.

Here, a coupled vibration model is considered to represent the interaction of two modes. These modes are assumed to oscillate with keeping the dynamics of power swing equations (3.14). The first vibration model corresponds to the most dominant mode, while the other corresponds to the second dominant mode. Three observation sites near generators that significantly participate in these two modes are selected; then, one end site is used as the reference of the phase angle. The dynamics of the model is represented by the polynomial approximation of the phase angle and the angular velocity:

$$\dot{x}_1 = \frac{\omega_r}{M_1}\{-D_1(\omega_1 - 1) + P_{m1} - P_{e1}\}$$

$$\approx f_1(x_1, x_2, x_3, x_4)$$

$$\dot{x}_2 = x_1$$

$$\dot{x}_3 = \frac{\omega_r}{M_2}\{-D_2(\omega_2 - 1) + P_{m2} - P_{e2}\} \qquad (6.1)$$

$$\approx f_2(x_1, x_2, x_3, x_4)$$

$$\dot{x}_4 = x_3$$

Here, functions f_1 and f_2 are assumed to consist of linear terms

$$\begin{bmatrix} \dot{x}_1 \\ \dot{x}_2 \\ \dot{x}_3 \\ \dot{x}_4 \end{bmatrix} = \begin{bmatrix} a_1 & a_2 & a_3 & a_4 \\ 1 & 0 & 0 & 0 \\ b_1 & b_2 & b_3 & b_4 \\ 0 & 0 & 1 & 0 \end{bmatrix} \begin{bmatrix} x_1 \\ x_2 \\ x_3 \\ x_4 \end{bmatrix} \qquad (6.2)$$

where $x_1 = \dot{\delta}_1 - \dot{\delta}_s, x_2 = \delta_1 - \delta_s - (\delta_{1e} - \delta_{se}), x_3 = \dot{\delta}_2 - \dot{\delta}_s$ and $x_4 = \delta_2 - \delta_s - (\delta_{2e} - \delta_{se})$. Subscripts 1 and 2 denote numbers of selected sites, the subscript "s" denotes the reference site, and the subscript "e" denotes the initial value of the phase angle at $t = 0$. The coefficients a_i and b_i $(i = 1 \sim 4)$ can be evaluated by applying the least squares method to time series data sets of x_j ($j = 1 \sim 4$) obtained by the wide area phasor measurement. Oscillation characteristics are investigated directly by using this model since eigenvalues of the coefficient matrix represent the damping and frequency of two oscillatory modes. If necessary, this model can be extended to the case expressed in more than two dominant modes, that is, by increasing state variables and choosing the corresponding measurement sites according to the number of the dominant modes.

On the other hand, the oscillation data obtained by the wide area measurement are the phase angle and include many frequency components associated with interarea low-frequency oscillations as well as local oscillations and many noises. Here, discrete wavelet analysis is applied to extract oscillations including the dominant two modes from measurement data. Interarea oscillations with frequencies between 0.2 and 0.8 Hz can be extracted, while modes with frequencies smaller than 0.2 Hz including DC components, local modes with frequencies higher than 1.0 Hz, and many noises with high frequencies can be eliminated by using wavelet transformation. Thus, data sets of phase difference x_2 and x_4 are provided. On the other hand, the angular velocity x_1 and x_3 can be calculated by differentiating x_2 and x_4, respectively.

Identified interarea low-frequency modes by using the mentioned method have occasionally poor damping characteristics. These modes, which are dominant in the wide-area stability, might become unstable with a heavier load on the tie-lines. Therefore, these modes must be damped by controlling the power system. The use of PSS is effective to stabilize interarea modes. Here, an approach for tuning of PSSs based on the phasor measurement is described.

Assume a dynamic model of the k-th PSS (with similar structure shown in Fig. 5.2) as follows,

$$G_{\mathrm{PSS}k}(s) = \frac{K_k}{1 + T_0 s} \cdot \frac{T_w s}{1 + T_w s} \cdot \frac{1 + T_{1k} s}{1 + T_{2k} s} \cdot \frac{1 + T_{3k} s}{1 + T_{4k} s} \tag{6.3}$$

which describes the typical real one consisting of a two-stage lead-lag compensation. Here, $T_{1k}, \ldots, T_{4k}$ are time constants, K_k is a gain, T_0 is time lag for the signal detection, and T_w is the time constant of the signal washout. In this study, PSS, which feeds back the generator angular velocity deviation $\Delta\omega$, is assumed, since a $\Delta\omega$-type PSS is more effective to damp interarea oscillations. When feeding back the generator output deviation ΔP, which is effective to damp local modes with frequency around 1.0 Hz, there is a necessity to carry out a synchronized measurement of the ΔP, or identify the ΔP from the measured phasors.

Here, the coupled vibration model (6.2) is extended for including the effect of PSSs:

$$\begin{bmatrix} \dot{x}_1 \\ \dot{x}_2 \\ \dot{x}_3 \\ \dot{x}_4 \\ \dot{x}_{\mathrm{PSS}1} \\ \dot{x}_{\mathrm{PSS}2} \end{bmatrix} = \begin{bmatrix} a_1' & a_2' & a_3' & a_4' & c_1^T \\ 1 & 0 & 0 & 0 & 0^T \\ b_1' & b_2' & b_3' & b_4' & c_2^T \\ 0 & 0 & 1 & 0 & 0^T \\ d_1 & 0 & 0 & 0 & D_1 \\ 0 & 0 & d_2 & 0 & D_2 \end{bmatrix} \begin{bmatrix} x_1 \\ x_2 \\ x_3 \\ x_4 \\ x_{\mathrm{PSS}1} \\ x_{\mathrm{PSS}2} \end{bmatrix} \tag{6.4}$$

where $x_{\mathrm{PSS}k}$ ($k = 1, 2$) is the vector of state variables of PSS defined by

$$x_{\mathrm{PSS}k} = \begin{bmatrix} v_{1k} & v_{2k} & v_{3k} & v_{4k} \end{bmatrix}^{\mathrm{T}}$$

$$v_{1k} \equiv \Delta V_{\mathrm{PSS}k}$$

(6.5)

where the superscript "T" stands for the transposition of vector. Variables x_2 and x_4 are directly measured by the PMUs. Other ones can be calculated by the relations of $x_1 = \dot{x}_2$, $x_3 = \dot{x}_4$, and (6.3) from the measured x_2 and x_4.

The vectors $\mathbf{c}_k$ ($k = 1, 2$), which are calculated simultaneously with a_i' and b_i' ($i = 1 \sim 4$) by the least squares method, imply the effect of PSS on each oscillation mode. Note that only one of the elements of $\mathbf{c}_k$ associated with the PSS output (v_{1k}) has any value, while all other elements are equal to zero, that is,

$$\mathbf{c}_1^{\mathrm{T}} = \begin{bmatrix} c_1 & 0 & 0 & 0 & 0 & 0 & 0 & 0 \end{bmatrix}$$

$$\mathbf{c}_2^{\mathrm{T}} = \begin{bmatrix} 0 & 0 & 0 & 0 & c_2 & 0 & 0 & 0 \end{bmatrix}$$

(6.6)

The 4×1 vector $\mathbf{d}_k$ and the 4×8 matrix $\mathbf{D}_k$ are determined by the real structure of object PSS (6.3), which contains the parameters K_k, T_0, T_{w}, and $T_{1k}, \ldots, T_{4k}$.

$$\mathbf{d}_k = \omega_r K_k \begin{bmatrix} \dfrac{T_{1k} T_{3k}}{T_{2k} T_{4k}} & 1 & 1 & \dfrac{T_{1k}}{T_{2k}} \end{bmatrix}^{\mathrm{T}}$$

(6.7)

$$\mathbf{D}_1 = \begin{bmatrix} \mathbf{DD}_1 & | & 0 \end{bmatrix}$$

$$\mathbf{D}_2 = \begin{bmatrix} 0 & | & \mathbf{DD}_2 \end{bmatrix}$$

$$\mathbf{DD}_k = \begin{bmatrix} -\dfrac{1}{T_0} & -\dfrac{T_{1k} T_{3k}}{T_{\mathrm{w}} T_{2k} T_{4k}} & \dfrac{T_{3k}}{T_{2k} T_{4k}}\left(1 - \dfrac{T_{1k}}{T_{2k}}\right) & \dfrac{1}{T_{4k}}\left(1 - \dfrac{T_{3k}}{T_{4k}}\right) \\[2ex] 0 & -\dfrac{1}{T_{\mathrm{w}}} & 0 & 0 \\[2ex] 0 & -\dfrac{1}{T_{\mathrm{w}}} & -\dfrac{1}{T_{2k}} & 0 \\[2ex] 0 & -\dfrac{T_{1k}}{T_{\mathrm{w}} T_{2k}} & \dfrac{1}{T_{2k}}\left(1 - \dfrac{T_{1k}}{T_{2k}}\right) & -\dfrac{1}{T_{4k}} \end{bmatrix}$$

(6.8)

Note that the input of each PSS in the model (6.4) is $\Delta\omega_1 = x_1/\omega_r$ and $\Delta\omega_2 = x_3/\omega_r$, respectively, although it is different from the real input signal of the $\Delta\omega$-type PSS. The model (6.4) should be considered to be a simplified model specialized for tuning of PSSs to search the appropriate direction of tuning parameters. The stability of dominant modes is evaluated by eigenvalues of matrix (6.4).

6.2.2 Tuning Mechanism

Two PSSs are tuned based on the model (6.4) since vectors $\mathbf{d}_k$ and matrices $\mathbf{D}_k$ include parameters of PSSs, and the change of parameters affect eigenvalues directly. Tuned parameter sets are expected to stabilize at least two modes although they could not be optimum sets. The effectiveness of the tuning approach can be assessed more properly by evaluating the eigenvalues of (6.4), which is obtained from measurement data after applying the tuned control method. Thus, the PSS parameters can be tuned based on the small-signal oscillation data obtained by the wide-area phasor measurement.

The obtained new PSS parameter sets by the mentioned approach are the final goal of the proposed tuning method. However, the obtained parameter sets might be affected by an error in the calculation process, a large noise in the measurement data, a sudden change of the system state, and the adverse effect on other modes that are not considered in the model (6.4). Therefore, the PSSs must be tuned by gradual steps for the security of the power system. This gradual tuning mechanism is also applicable in the case of using other methods for tuning of PSS parameters, such as identification-based methods, genetic algorithm, fuzzy logic, and so on.

The wide-area phasor measurement-based PSS tuning mechanism using the described extended coupled vibration model (6.4) consists of the following steps:

Step 1: Select three measurement sites, which significantly participate in the dominant and the quasi-dominant modes. One of them is for the reference of the phase angle and the other two are targets of the tuning issue.

Step 2: Obtain the oscillation data from the sites. Steady state phasor fluctuations measured in the normal operation are available. Note that phasors with time labels are stored using the installed PMUs (synchronized by the GPS signal). Therefore, measured phasors are directly comparable with each other without considering the communication delay.

Step 3: Extract interarea oscillations by applying the discrete wavelet transformation to the observed phasor. The time series data sets of the PSS state variables can be provided by calculating (6.3) in the time domain.

Step 4: Determine coefficients a_i', b_i', and c_k of the coupled vibration model (6.4) by the least squares method.

Step 5: Find appropriate parameter sets of PSSs based on the model (6.4).

Step 6: Tune the PSSs step by step toward the parameter sets obtained in the previous step. In each step, the effect of tuning is assessed by eigenvalues of (6.4) that is modeled again using the phasor fluctuations after tuning control.

Step 7: By repeating the gradual tuning and the stability assessment, the PSSs can be tuned appropriately.

Here, a coupled vibration model with two dominant electro-mechanical modes associated with the angle stability is discussed. Therefore, this approach is applicable in the case that a dominant and a quasi-dominant mode are classified while the participating measurement sites (in these modes) are specified. Note that the application of this

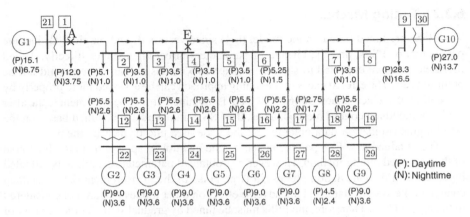

Figure 6.2. IEEJ WEST 10-machine system model.

approach is not limited to the power system with longitudinal configuration used in this study.

6.2.3 Simulation Results

The method is applied to a 10-machine longitudinally interconnected system. Figure 6.2 shows the configuration of the IEEJ WEST 10-machine system model. The detailed parameters and relevant data are available in Ref. [10]. This test system represents the power system model of the western 60 Hz areas of Japan. System constants and generation capacity of generators are shown in Tables 2.1 and 6.1, respectively. The active power of all loads is assumed to have constant current characteristics, and reactive power is considered with constant impedance characteristics. In Fig. 6.2, (P) denotes the generation and loading condition in the daytime and (N) denotes the condition in the nighttime. The unit is 1000 MVA base per unit.

Each generator is equipped with an automatic voltage regulator (AVR) shown in Fig. 6.3, and generators 1 and 5 are equipped with the $\Delta\omega$-type PSS to damp the interarea

TABLE 6.1. Generator Rated Capacity (MVA)

	Daytime, MVA	Nighttime, MVA
G1	15,000	9,000
G8	5,000	3,000
G10	30,000	18,000
Others	10,000	6,000
Total sum	120,000	72,000

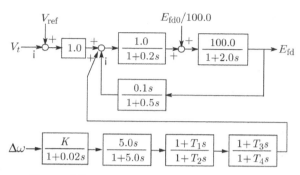

Figure 6.3. Exciter model with the $\Delta\omega$ type PSS.

oscillations. In this study, the time lag for the signal detection and the time constant of the signal washout are $T_0 = 0.02$ (s), $T_w = 5.0$ (s), respectively. In the initial/nominal condition, the parameter setting of PSS is $K = 0.20$, $T_1 = 1.66$, $T_2 = 1.51$, $T_3 = 2.07$, $T_4 = 0.01$, and two generators are equipped with the identical PSS. In this study, the EUROSTAG software [11] is used for the time domain simulation of the power system dynamics.

Figure 2.3 shows the real part of the right eigenvectors corresponding to the generator rotor angle in the daytime loading condition. Mode 1 denotes the dominant mode, and mode 2 denotes the quasi-dominant mode. This result shows that mode 1 oscillates in the opposite direction between both end generators 1 and 10, while mode 2 oscillates between both end generators 1, 10, and the middle generator 5. Figure 2.4 shows participation factors corresponding to the generator rotor angle. These results show that generators 1, 5, and 10 participate principally in modes 1 and 2. Here, the coupled vibration model is calculated by using the voltage phasor of nodes 21 and 25 with node 30 as a reference.

In this study, some small load variations at some load buses are assumed to simulate the phasor fluctuations measured in the real power system. The extended coupled vibration model (6.4) is constructed by using these small oscillations. Table 6.2 shows the eigenvalues variation when each parameter of the connected PSS to the generator 1 is changed. The second column of the table shows eigenvalues of dominant low-frequency modes calculated by using the power system simulation package. The third column shows eigenvalues of the corresponding two modes of the approximate model (6.4). The result shows that the approximate model estimates two conjugate eigenvalues, properly. In addition, the trend of the variation of both eigenvalues with parameter change coincides well with each other. Therefore, the extended coupled vibration model can correctly evaluate the stability change with tuning of PSS parameters. The advantage of this approach is in estimating the dominant modes by using steady state phasor fluctuations measured in the normal operating condition.

Table 6.3 shows the result of tuning of PSSs based on the approximate model (6.4). The PSS parameter sets are determined to improve damping ratios of two low-frequency

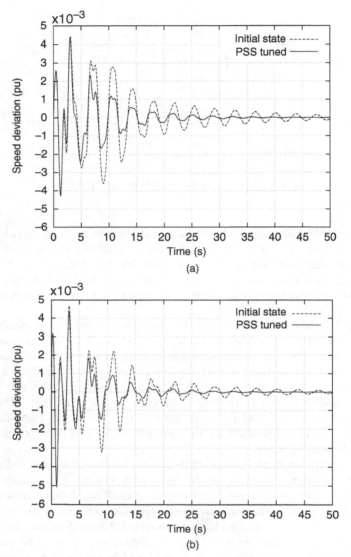

Figure 6.4. Simulation with the tuned PSSs of generators 1 and 5 in the daytime condition. (a) Generator 1. (b) Generator 5.

modes of the model, that is, to maximize the following objective function J under the constraint that all eigenvalues lie on the left-hand side of the complex plane:

$$\max J = \min_{i} \frac{-\mathrm{Re}(\lambda_i)}{\sqrt{\mathrm{Re}(\lambda_i)^2 + \mathrm{Im}(\lambda_i)^2}} \tag{6.9}$$

TABLE 6.2. The Comparison of Eigenvalues Variation of the Original System and the Approximate Model, When the Parameters of the Connected PSS to Generator 1 are Changed in the Daytime Loading Condition

Parameters	Full System	Approximate Model
Initial condition	$-0.065 \pm j1.731$	$-0.080 \pm j2.223$
	$-0.154 \pm j4.055$	$-0.182 \pm j3.832$
$K = 0.02$	$-0.048 \pm j1.714$	$-0.074 \pm j2.152$
	$-0.152 \pm j4.050$	$-0.179 \pm j3.815$
$K = 0.4$	$-0.080 \pm j1.751$	$-0.085 \pm j2.292$
	$-0.157 \pm j4.061$	$-0.184 \pm j3.847$
$T_1 = 2.16$	$-0.071 \pm j1.735$	$-0.082 \pm j2.248$
	$-0.155 \pm j4.057$	$-0.183 \pm j3.835$
$T_1 = 1.16$	$-0.058 \pm j1.728$	$-0.078 \pm j2.196$
	$-0.153 \pm j4.054$	$-0.181 \pm j3.828$
$T_2 = 2.01$	$-0.059 \pm j1.728$	$-0.078 \pm j2.201$
	$-0.154 \pm j4.054$	$-0.181 \pm j3.828$
$T_2 = 1.01$	$-0.076 \pm j1.736$	$-0.083 \pm j2.261$
	$-0.156 \pm j4.058$	$-0.183 \pm j3.836$
$T_3 = 2.57$	$-0.070 \pm j1.735$	$-0.082 \pm j2.243$
	$-0.155 \pm j4.057$	$-0.183 \pm j3.835$
$T_3 = 1.57$	$-0.060 \pm j1.728$	$-0.078 \pm j2.202$
	$-0.154 \pm j4.054$	$-0.181 \pm j3.828$
$T_4 = 0.51$	$-0.048 \pm j1.732$	$-0.075 \pm j2.154$
	$-0.150 \pm j4.052$	$-0.179 \pm j3.826$

subject to

$$\text{Re}(\lambda_i) < 0 \quad (i = 1, \ldots, n) \tag{6.10}$$

where λ_i is the i-th eigenvalue of the model (6.4). The $\text{Re}(\lambda_i)$ and $\text{Im}(\lambda_i)$ denotes the real and imaginary parts of the i-th eigenvalue, respectively. The optimization is implemented by using the MATLAB® Toolbox for genetic algorithm [12].

Table 6.4 shows parameter settings for five steps of gradual tuning of the PSS connected to generator 1. One wrong case with the opposite direction of parameter tuning

TABLE 6.3. Result of the PSS Tuning in the Daytime Condition

Parameter	Initial	G1	G5
K	0.20	0.37	0.23
T_1	1.66	1.82	2.64
T_2	1.51	0.83	0.64
T_3	2.07	2.31	2.09
T_4	0.01	0.01	0.01

TABLE 6.4. The Stability Change with Gradual Tuning of PSS Connected to Generator 1

Parameter Sets	K	T_1	T_2	T_3	T_4	Eigenvalues of the Approximate Model	Damping Ratio, %
Wrong	0.16	1.63	1.66	2.02	0.01	$-0.0781 \pm j2.1976$	3.554
						$-0.1808 \pm j3.8266$	4.720
Initial	0.20	1.66	1.51	2.07	0.01	$-0.0801 \pm j2.2230$	3.601
						$-0.1816 \pm j3.8316$	4.735
First	0.24	1.69	1.36	2.12	0.01	$-0.0822 \pm j2.2521$	3.646
						$-0.1828 \pm j3.8369$	4.759
Second	0.28	1.72	1.21	2.17	0.01	$-0.0845 \pm j2.2868$	3.691
						$-0.1841 \pm j3.8426$	4.786
Third	0.32	1.75	1.06	2.22	0.01	$-0.0866 \pm j2.3250$	3.723
						$-0.1854 \pm j3.8478$	4.812
Fourth	0.36	1.78	0.91	2.27	0.01	$-0.0886 \pm j2.3644$	3.744
						$-0.1868 \pm j3.8522$	4.843
Final	0.37	1.82	0.83	2.31	0.01	$-0.0894 \pm j2.3836$	3.749
						$-0.1876 \pm j3.8531$	4.862

is also considered for investigating the accuracy of the proposed method. Table 6.4 also shows eigenvalues of the coupled vibration model (6.4) and damping ratios in each parameter setting. The result shows that two dominant modes are stabilized gradually with the parameter change from the initial parameter set to the final set. On the other hand, in the wrong case, two modes are destabilized with the change of parameters. In case of considering the connected PSS to generator 5, similar results have been obtained. These results show that appropriate parameters of PSSs can be obtained and the stability change can be exactly assessed by using the coupled vibration model.

Figure 6.4 shows the simulation results when two tuned PSSs are applied, simultaneously. Here, a three-phase ground fault at point E in the double circuit transmission line, which is shown in Fig. 6.2, is assumed as a disturbance. The fault is cleared at 0.07 s after the fault occurred. The result shows that low-frequency oscillation is effectively damped by the tuned PSSs. On the other hand, the critical clearing time, which was 0.091 s in the initial condition, is improved up to 0.142 s with the tuned PSSs.

The proposed method is applied to a different loading condition. The loading condition denoted by (N) in Fig. 6.2 is used in this analysis. Table 6.5 shows the comparison of eigenvalues calculated by using the full system matrix with the approximate coupled vibration model. The result shows that the approximate model successfully evaluates the slight change of eigenvalues.

Table 6.6 and Fig. 6.5 show the results of tuning the PSSs. Assumed disturbance is the same fault at point A in Fig. 6.2, where it is cleared at 0.07 s after the fault occurred. The critical clearing time is improved from 0.070 s up to 0.087 s. These results demonstrate the effectiveness of the proposed approach based on the wide-area phasor measurement.

TABLE 6.5. The Comparison of Eigenvalues Variation of the Original System and the Approximate Model, When the Parameters of the Connected PSS to Generator 1 is Changed in the Nighttime Loading Condition

Parameters	Full System	Approximate Model
Initial condition	$-0.1021 \pm j2.7001$	$-0.0875 \pm j2.7733$
	$-0.1912 \pm j5.0542$	$-0.2346 \pm j4.3954$
$K = 0.02$	$-0.0996 \pm j2.6966$	$-0.0872 \pm j2.7699$
	$-0.1902 \pm j5.0497$	$-0.2341 \pm j4.3900$
$K = 0.4$	$-0.1049 \pm j2.7040$	$-0.0878 \pm j2.7771$
	$-0.1924 \pm j5.0591$	$-0.2358 \pm j4.4002$
$T_1 = 2.16$	$-0.1032 \pm j2.7011$	$-0.0878 \pm j2.7747$
	$-0.1918 \pm j5.0557$	$-0.2352 \pm j4.3969$
$T_1 = 1.16$	$-0.1010 \pm j2.6992$	$-0.0870 \pm j2.7716$
	$-0.1907 \pm j5.0527$	$-0.2344 \pm j4.3930$
$T_2 = 2.01$	$-0.1013 \pm j2.6993$	$-0.0873 \pm j2.7724$
	$-0.1908 \pm j5.0530$	$-0.2347 \pm j4.3929$
$T_2 = 1.01$	$-0.1040 \pm j2.7014$	$-0.0888 \pm j2.7765$
	$-0.1923 \pm j5.0565$	$-0.2354 \pm j4.3979$
$T_3 = 2.57$	$-0.1029 \pm j2.7010$	$-0.0876 \pm j2.7744$
	$-0.1916 \pm j5.0554$	$-0.2351 \pm j4.3965$
$T_3 = 1.57$	$-0.1013 \pm j2.6993$	$-0.0873 \pm j2.7725$
	$-0.1908 \pm j5.0530$	$-0.2347 \pm j4.3931$
$T_4 = 0.51$	$-0.0984 \pm j2.6990$	$-0.0651 \pm j2.7355$
	$-0.1885 \pm j5.0503$	$-0.2300 \pm j4.3518$

TABLE 6.6. The Result of PSSs Tuning in the Nighttime Condition

Parameters	Initial	G1	G5
K	0.20	0.31	0.67
T_1	1.66	4.52	1.91
T_2	1.51	0.68	0.43
T_3	2.07	3.58	2.98
T_4	0.01	0.01	0.01

6.3 WIDE-AREA MEASUREMENT-BASED CONTROLLER DESIGN

6.3.1 Wide-Area Power System Identification

This section introduces another approach for tuning of the damping controller with wide-area phasor measurements. The interarea oscillation dynamics with a single mode can be simplified by assuming that it has an analogy of a single machine and infinite bus system:

$$\dot{\omega} = f(\Delta\omega, \Delta\delta)$$
$$\dot{\delta} = \omega \tag{6.11}$$

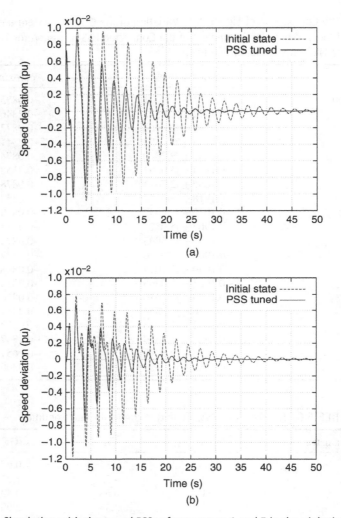

Figure 6.5. Simulation with the tuned PSSs of generators 1 and 5 in the nighttime condition. (a) Generator 1. (b) Generator 5.

Using voltage phasors of two measurement sites, a second-order oscillation model for the dominant oscillation mode can be identified in the following form:

$$
\begin{bmatrix} \dot{x}_1 \\ \dot{x}_2 \end{bmatrix} = \begin{bmatrix} a_1 & a_2 \\ a & 0 \end{bmatrix} \begin{bmatrix} x_1 \\ x_2 \end{bmatrix} = \mathbf{A} \begin{bmatrix} x_1 \\ x_2 \end{bmatrix}
\tag{6.12}
$$

where $x_1 = \dot{\delta}_1 - \dot{\delta}_s$, and $x_2 = \delta_1 - \delta_s - (\delta_{1e} - \delta_{se})$. The subscript 1 denotes the selected site, the subscript "s" denotes the reference site, and subscript "e" denotes the initial

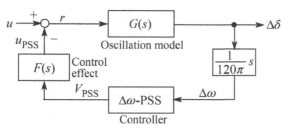

Figure 6.6. Approximate low-order model for tuning the $\Delta\omega$-type PSS to damp inter-area low-frequency oscillation.

phase angle at $t=0$. The coefficients a_1 and a_2 can be determined by the least squares method by using filtered voltage phasors data with the fast Fourier transform (FFT)-based filtering method described in Chapter 2. The characteristics of the extracted mode can be evaluated by the eigenvalues of the coefficient matrix $\mathbf{A}$. Assuming that the eigenvalues are given by $\alpha \pm j\beta$, the approximate oscillation model of the interested mode is represented by the following form:

$$G(s) = \frac{1}{s^2 - 2\alpha s + \left(\alpha^2 + \beta^2\right)} \tag{6.13}$$

Considering a damping controller design, the control loop with the oscillation model $G(s)$ and a controller is assumed as shown in Fig. 6.6. Here, the damping controller, where a PSS is considered in this study, includes the real structure of the controller, and the model $F(s)$ represents the effect of the controller on the oscillation mode $G(s)$. In order to simplify the procedure of the controller design, the order of $F(s)$ should be reduced as much as possible. In this study, the model $F(s)$ is assumed by the following form:

$$F(s) = K \cdot \frac{(1 + T_a s)(1 + T_b s)}{(1 + T_c s)(1 + T_d s)(1 + T_e s)} \tag{6.14}$$

The procedure of determining the model $F(s)$ is as follows:

Step 1: Determine time constants $T_a \sim T_d$ to maximize the gain characteristics around the frequency, which corresponds to the imaginary part of the estimated eigenvalue (β).

Step 2: Set the time constant T_e at about zero degree of the frequency response of $F(s)$ around the frequency, which corresponds to the imaginary part β of the estimated eigenvalue.

Step 3: Compute the gain K to minimize the mean square error between the filtered phase difference and the output of the approximate model shown in Fig. 6.6.

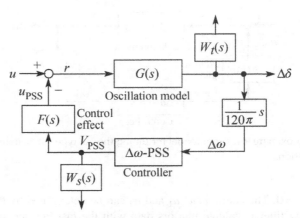

Figure 6.7. Closed-loop system configuration.

6.3.2 Design Procedure

Here, the H_∞ control theory is applied to the approximate low-order system model shown in Fig. 6.6 to compensate for the modeling error. Figure 6.7 shows the generalized plant including two weighting functions, $W_s(s)$ and $W_t(s)$. The $W_s(s)$ has low-pass filter characteristics to consider the robust stability and modeling error, while $W_t(s)$ has high-pass filter characteristics to consider the influence of other modes. The H_∞ norm of the closed-loop system can be given by

$$\left\| \begin{matrix} W_s(s)\dfrac{P(s)T_g(s)}{1+P(s)T_g(s)} \\ W_t(s)\dfrac{P(s)T_g(s)}{1+P(s)T_g(s)} \end{matrix} \right\|_\infty < \gamma \tag{6.15}$$

where $P(s)$ is the transfer function of the controller. $T_g(s)$ is the transfer function of the generalized plant including the simplified model $G(s)$, $F(s)$, and scaling factor. An H_∞ controller satisfying (6.15) is designed for the minimum γ, as the H_∞ robustness index.

Note that the designed controller generally has a high order; therefore, the order of the controller should be as low as possible to simplify the tuning process. In this study, the balanced realization technique is adopted to obtain a low-order controller from the original designed H_∞ controller. It is noteworthy that instead of H_∞ control, other robust/advanced control techniques such as H_2 and μ control synthesis methods can be also used [13].

6.3.3 Simulation Results

The method is applied to the simple 11-bus two-area power system model shown in Fig. 6.8. System constants including detailed bus and line data as well as the generators and load parameters are given in Ref. [14].

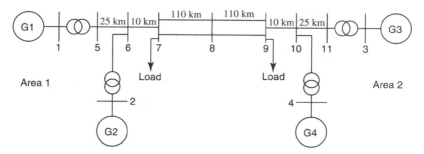

Figure 6.8. Simple two-area system model.

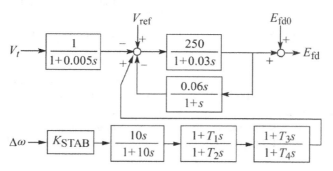

Figure 6.9. Exciter and PSS model for generator 1.

Generator 1 is equipped with an AVR and a $\Delta\omega$-type PSS shown in Fig. 6.9 to damp interarea oscillations. PSS parameters for the initial condition are $K_{STAB} = 5.0$, $T_1 = 2.0$, $T_2 = 1.0$, $T_3 = 2.0$, and $T_4 = 0.5$. Here, K_{STAB} is the PSS gain. Other generators are equipped with the identical AVR with PSS shown in Fig. 6.10. The effect of the speed governor is not considered. In this study, the EUROSTAG software [11] has been used for the simulation of the power system dynamics.

First, the system is examined without connecting the PSS of generator 1. The small load fluctuations at node 7 are assumed to simulate the phasor fluctuations measured in

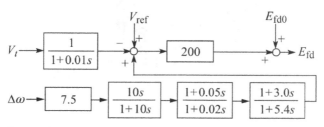

Figure 6.10. Exciter and PSS model for generators 2, 3, and 4.

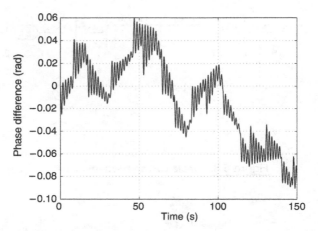

Figure 6.11. Phase difference between nodes 1 and 3.

the normal condition of the real power system. Figure 6.11 shows the phase differences between nodes 1 and 3. Figure 6.12 shows the Fourier spectrum of Fig. 6.11. The result shows that an interarea oscillation mode with frequency of around 0.60 Hz is dominant.

Figure 6.13 shows the phase difference of interarea low-frequency oscillations extracted by the FFT-based filter, where the frequency band width is set to

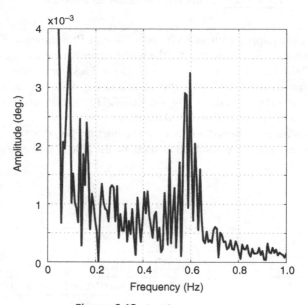

Figure 6.12. Fourier spectrum.

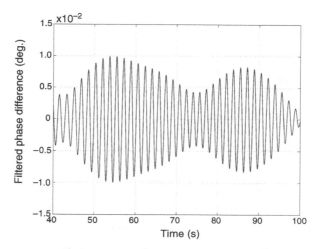

Figure 6.13. Filtered phase difference.

$0.60 \pm 0.04\,\mathrm{Hz}$. A simplified oscillation model (6.12) is identified by applying the least squares method to the filtered phase difference data. The eigenvalues of the coefficient matrix **A** are $\alpha \pm j\beta = -0.137 \pm j3.720$. Substituting the obtained values into (6.12), the oscillation model (6.13) is represented as follows:

$$G(s) = \frac{1}{s^2 + 0.274s + 13.86} \tag{6.16}$$

The model $F(s)$ is identified by using the procedure described in Section 6.3.1. The model $F(s)$ is given by

$$F(s) = 0.87 \cdot \frac{(1 + 0.82s)(1 + 0.58s)}{(1 + 0.45s)(1 + 0.31s)(1 + 0.22s)} \tag{6.17}$$

Figure 6.14 illustrates the bode diagram of the model $F(s)$, which has the gain peak at $3.77\,\mathrm{rad/s}$.

The input u of the approximate model shown in Fig. 6.6 is estimated by multiplying the inverse function of (6.16) to filtered phase difference data. Figure 6.15 shows the estimated input signal u of the approximate model. Figure 6.16 shows the comparison between the filtered phase difference of measured data and the approximate model output simulated by using the estimated input signal. These results demonstrate that the approximate model can successfully represent the characteristics of the interarea low-frequency oscillation mode and the effect of the controller on this mode.

The H_∞ controller is designed for the identified low-order system model. The controller corresponding to the minimum γ satisfying the constraint (6.15) is derived

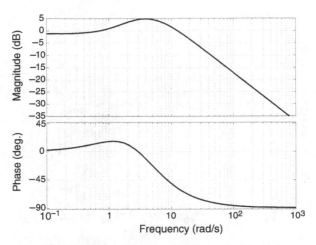

Figure 6.14. Bode diagram of the model $F(s)$.

based on the generalized plant shown in Fig. 6.7. Two weighting functions $W_s(s)$ and $W_t(s)$ are given by

$$
\begin{aligned}
W_s(s) &= 15 \cdot \frac{1 + 0.0003s}{1 + 0.05s} \\
W_t(s) &= 0.1 \cdot \frac{1 + 10s}{1 + 0.01s}
\end{aligned}
\tag{6.18}
$$

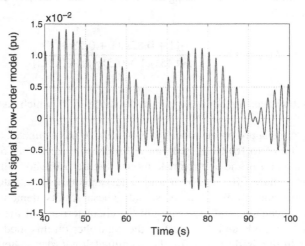

Figure 6.15. The input signal u for the low-order model.

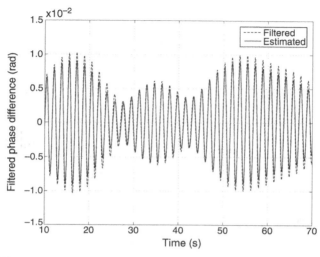

Figure 6.16. The comparison between the filtered phase difference and the output signal of the low-order model.

The H_∞ controller is derived in the following form:

$$P_g(s) = \frac{a_7 s^7 + a_6 s^6 + \cdots + a_1 s + a_0}{s^8 + b_7 s^7 + b_6 s^6 + \cdots + b_1 s + b_0} \tag{6.19}$$

The coefficients of the H_∞ controller (6.19) are shown in Table 6.7.

By applying the balanced realization technique to the H_∞ controller (6.19), the final model of reduced order PSS is obtained as follows:

$$P(s) = \frac{1.39 \times 10^6 s^2 + 3.05 \times 10^7 s + 1.48 \times 10^6}{s^3 + 3068 s^2 + 1.89 \times 10^5 s + 5.47 \times 10^4} \tag{6.20}$$

TABLE 6.7. The Coefficients of H_∞ Controller

a_7	1.39×10^6	b_7	3161
a_6	1.64×10^8	b_6	4.88×10^5
a_5	4.03×10^9	b_5	2.02×10^7
a_4	3.03×10^{10}	b_4	1.85×10^8
a_3	8.94×10^{10}	b_3	5.93×10^8
a_2	9.46×10^{10}	b_2	7.10×10^8
a_1	1.20×10^{10}	b_1	2.06×10^7
a_0	3.43×10^8	b_0	1.40×10^7

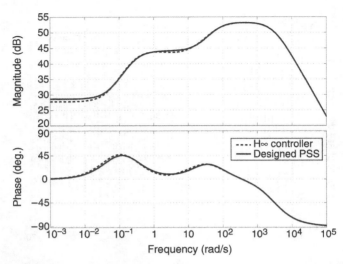

Figure 6.17. The comparison of bode diagrams between the full-order (6.19) and the reduced-order (6.20) H_∞ controllers.

Figure 6.17 demonstrates a comparison of bode diagrams of the H_∞ controller (6.19) and the designed PSS (6.20). The result shows that the designed PSS (6.20) with third order is the same in performance with the H_∞ controller (6.19) with eighth order.

Here, the performance of the designed PSS (6.20) by the proposed method is evaluated. Table 6.8 shows eigenvalues of the interarea low-frequency mode and two local modes of the original system model shown in Fig. 6.8. Appropriate parameters of the PSS can be obtained since the interarea mode is damped effectively. Note that the low-order model does not consider other modes than the interarea mode; therefore, it is difficult to evaluate the influence of the parameter tuning of the controller on other modes on the low-order model. In this case, the influence is not so large as shown in Table 6.8 since local modes have larger damping than the interarea mode.

Figure 6.18 shows the step response of the approximate model, where the magnitude of the step input has been adjusted to be consistent with the response of the original system shown in Fig. 6.19. The result shows that the designed PSS (6.20) can damp the low-frequency oscillation more effectively than the initial condition.

On the other hand, Fig. 6.19 shows the simulation results of the original system shown in Fig. 6.8. An assumed disturbance here is a three-phase to ground fault between

TABLE 6.8. The Eigenvalues of the Simple Two-Area System

Mode	Interarea	Area 1 Local	Area 2 Local
Without PSS	$-0.123 \pm j3.710$	$-0.958 \pm j6.863$	$-1.176 \pm j7.763$
Initial PSS	$-0.201 \pm j3.814$	$-0.916 \pm j7.243$	$-1.175 \pm j7.764$
Designed PSS	$-0.322 \pm j4.071$	$-0.830 \pm j8.743$	$-1.173 \pm j7.760$

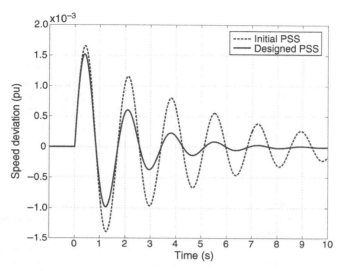

Figure 6.18. Step response with the low-order model.

nodes 7 and 8. The fault is cleared at 0.067 s after the fault. The result shows that the interarea oscillation is effectively damped by the designed PSS.

Comparing Fig. 6.18 with Fig. 6.19, characteristics of the interarea low-frequency oscillation of the original system can be represented approximately by the low-order model. These results demonstrate the effectiveness of the proposed method based on wide-area phasor measurements considering the error compensation with H_∞ control theory.

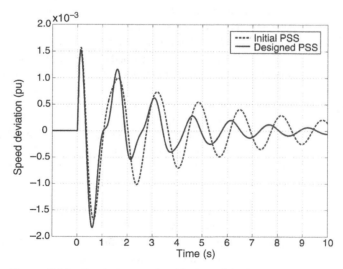

Figure 6.19. Simulation result with the original two-area system.

6.4 SUMMARY

This chapter describes a method for tuning of PSSs based on wide-area phasor measurements. The low-order system model, which holds the characteristics of the interarea oscillation mode and the control effect, is identified by monitoring data from wide-area phasor measurements. The effectiveness of the proposed method has been demonstrated through the power system simulation. The results show that an appropriate controller can be designed by using the identified low-order model.

REFERENCES

1. T. Michigami, Development of a new two-input PSS to control low-frequency oscillation in interconnecting power systems and the study of a low-frequency oscillation model, *Electr. Eng. Jpn.*, **115**(7), 49–68, 1995.

2. T. Senjyu, T. Yamashita, and K. Uezato, Stabilization control of multi-machine power systems by adaptive power system stabilizer using frequency domain analysis, *Electr. Eng. Jpn.*, **142**(2), 10–20, 2003.

3. M. Ishimaru, R. Yokoyama, O. M. Neto, and K. Y. Lee, Allocation and design of power system stabilizers for mitigating low-frequency oscillations in the eastern interconnected power system in Japan, *Int. J. Electr. Power Energy Syst.*, **26**(8), 607–618, 2004.

4. A. G. Phadke, Synchronized phasor measurements in power systems, *IEEE Comput. Appl. Power*, **6**(2), 10–15, 1993.

5. T. Hashiguchi, M. Watanabe, A. Matsushita, Y. Mitani, O. Saeki, K. Tsuji, M. Hojo, and H. Ukai, Identification of characterization factor for power system oscillation based on multiple synchronized phasor measurements, *Electr. Eng. Jpn.*, **163**(3), 10–18, 2008.

6. D. J. Trudnowski, J. R. Smith, T. A. Short, and D. A. Pierre, An application of Prony methods in PSS design for multimachine systems, *IEEE Trans. Power Syst.*, **6**(1), 118–126, 1991.

7. I. Kamwa, G. Trudel, and L. Gerin-Lajoie, Low-order black-box models for control system design in large power systems, *IEEE Trans. Power Syst.*, **11**(1), 303–311, 1996.

8. A. Hasanovic, A. Feliachi, A. Hasanovic, N. B. Bhatt, and A. G. DeGroff, Practical robust PSS design through identification of low-order transfer functions, *IEEE Trans. Power Syst.*, **19**(3), 1492–1500, 2004.

9. N. Kakimoto, A. Nakanishi, and K. Tomiyama, Instability of interarea oscillation mode by autoparametric resonance, *IEEE Trans. Power Syst.*, **19**(4), 1961–1970, 2004.

10. Technical Committee of IEEJ, Japanese Power System Models, 1999. Available at http://www2.iee.or.jp/ver2/pes/23-st_model/english/index.html.

11. M. Stubbe, A. Bihain, J. Deuse, and J. C. Baader, STAG: a new unified software program for the study of the dynamic behaviour of electrical power systems, *IEEE Trans. Power Syst.*, **4**(1), 129–138, 1989.

12. *GAOT: A Genetic Algorithm for Function Optimization: A Matlab Implementation*, 1995. Available at http://www.ie.ncsu.edu/mirage/GAToolBox/gaot/.

13. H. Bevrani, *Robust Power System Frequency Control*, Springer, New York, 2009.

14. P. Kundur, Power System Stability and Control, McGraw-Hill, New York, 1994.

7

COORDINATED DYNAMIC STABILITY AND VOLTAGE REGULATION

Power systems continuously experience changes in operating conditions due to variations in generation/load and a wide range of disturbances. Power system stability and voltage regulation have been considered as important control problems for secure system operation over the years. Currently, because of expanding physical setups, functionality, and complexity of power systems, the proper design of automatic voltage regulators (AVRs) and power system stabilizers (PSSs) has becomes more significant than in the past. That is why in recent years a great deal of attention has been paid to the application of advanced control techniques for power systems.

Conventionally, the AVR and PSS design is considered as a sequential design including two separate stages without any coordination. Although it is known that the stability and voltage regulation issues are ascribed to different model descriptions, it has been long recognized that the AVR and PSS have inherent conflicting objectives.

In the present chapter, the necessity for coordination of AVR and PSS designs is emphasized, and three synthesis methodologies to enhance the stability and voltage regulation of existing real power systems without opening their conventional PSS and AVR loops are introduced.

Power System Monitoring and Control, First Edition. Hassan Bevrani, Masayuki Watanabe, and Yasunori Mitani.

7.1 NEED FOR AVR–PSS COORDINATION

As mentioned, the AVR and PSS are conventionally designed as two separate stages in the classical sequential synthesis procedure. First, the AVR is designed to meet the specified voltage regulation performance and then the PSS is designed to satisfy the stability and required damping performance. The conventional PSS and AVR structures are described in Chapter 5.

An AVR keeps the generator terminal voltage at a preset value, and improves transient stability. However, the fast responding AVRs deteriorate small signal stability by introducing electromechanical modes in the power system [1]. Continuous load/generation changes and various disturbances impose low frequency oscillations in a power system [2]. When an electromechanical oscillation occurs, the torque resolved into two components, one in phase with machine rotor angle (synchronizing torque) and the other in phase with machine rotor speed (damping torque). The lack of synchronizing and/or damping torque may lead to system instability. Before the widespread use of AVRs, instability has mainly occurred due to lack of synchronizing torque. This type of instability was manifested in the form of aperiodic drift of the rotor angle of the synchronous machines. The installed AVRs improve synchronizing torque in the power system. Other types of instability are the result of the lack of damping torque as sustained or increased oscillations of rotor angles [3]. Therefore, the AVRs improve transient stability due to the increase in the synchronizing torque between interconnected generators, and the decrease in small-signal stability and damping torque due to the increase of rotor oscillations. A conventional AVR and excitation control system is shown in Fig. 7.1a.

The PSSs are used to attenuate the low-frequency oscillations. The PSSs are employed to produce an auxiliary damping torque and enhance the stability margin. A PSS usually includes two phase-lead compensators (to compensate existing phase-lag

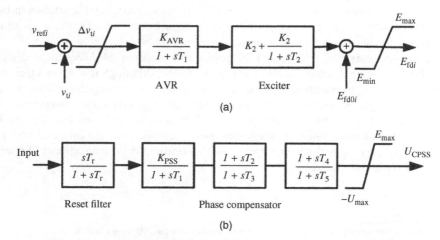

Figure 7.1. (a) Conventional AVR and excitation control system. (b) Conventional PSS.

between exciter input and electrical torque), a reset/washout filter (to eliminate DC signals), and a constant gain as shown in Fig. 7.1b. The AVR and PSS units are usually connected together as shown in Fig. 5.2.

Rotors of generating units such as steam generator turbines are made of several connected pieces through the shaft with different masses. Although in power system dynamic analysis, the steam generator–turbine rotor is usually considered as a single mass unit; however, a disturbance may practically create torsional oscillations between the pieces as well as low-frequency oscillations. This problem to be magnified concerning the existing installed PSSs [4]. Usually to remove torsional oscillation, torsional filters are required to provide appropriate input speed/frequency signals for the PSSs.

The AVRs and PSSs produce torques in phase with the rotor angle and speed variations, respectively. However, both AVR and PSS employ field voltage to produce different rotor angle and speed-based torques, but an enhancement in one direction may cause deterioration in the other direction. Therefore, due to this conflict, a tradeoff between AVR and PSS control actions is required. The impact of AVR and PSS on the power system stability is shown in Fig. 7.2.

In this figure, the torques resolved into damping and synchronizing components. The system is stable when these components have positive values. In Fig. 7.2a, the impact of constant excitation without PSS and AVR on the power system is depicted. It clarifies that the system is operating in a stable condition. Adding AVR injects an extra torque with positive synchronizing and large negative damping component to the system, which result in total negative damping torque (Fig. 7.2b). The lack of damping torque component in this condition makes the system unstable (oscillatory instability). Applying a torque in phase with the rotor speed variation (using PSS) compensates the lack of damping torque and conduct the system to a stable condition (Fig. 7.2c).

Numerous oscillation modes exist in a large-scale power system, wherein removing all of them is neither practical nor economical. However, there are two important modes, namely, local and interarea modes that should be controlled by the PSSs. The local mode appears in the frequency range of 0.8–2.0 Hz and occur when generators in a plant swing against the rest of the power system. As described in Chapters 2 and 6, the interarea mode (in the range of 0.1–0.7 Hz) occurs when two groups of generators in different areas swing against each other.

The main issue in using of AVR and PSS is how to tune their parameters. In general, the PSS philosophy relies on phase compensation using linear control theory. The PSS parameters are tuned to provide a desirable performance at the nominal operating point. Most of the previous works considered the conventional AVR and PSS structures with two separate design stages. First, the AVR parameters are tuned to achieve acceptable transient stability, and then the PSS is designed to meet the required damping characteristic.

Although, it has been long recognized that AVR and PSS have inherent conflicting objectives, most of the previous works have addressed the AVR and PSS designs, neglecting the existing conflict. Furthermore, so many simplified assumptions in these reports degrade the performance of controllers in practice [1].

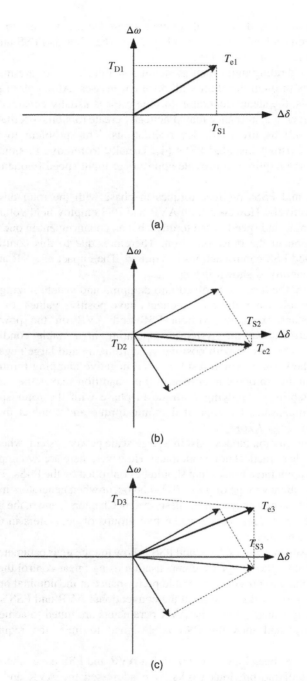

Figure 7.2. Impact of AVR and PSS on the power system stability: (a) constant excitation, (b) constant excitation and AVR, and (c) constant excitation with AVR and PSS.

In the last two decades, several reports considering an integrated synthesis approach for the AVR and PSS designs and covering various requirements for stabilization and voltage regulation within unique control structure have been published. Although most of the addressed approaches have been proposed based on new contributions in modern control systems, because of complexity of the control structure, numerous unknown design parameters, neglecting real constraints, and probable risk of placement instead of well-known conventional AVR/PSS units, they are not well suited to meet the design objectives in real multimachine power systems.

In response to these problems, this chapter presents three tuning mechanisms to enhance the stability and voltage regulation of existing real power systems without removing conventional PSS and AVR units. The proposed coordination methodologies use measurable signals for robust tuning of control proportional gains; therefore, they have considerable promise for implementation, especially in a large-scale multimachine system. In fact, the proposed control strategies attempt to make a bridge between the simplicity of control structure and robustness of stability and performance to satisfy simultaneous AVR and PSS objectives.

7.2 A SURVEY ON RECENT ACHIEVEMENTS

In the last few decades, some studies have considered an integrated design approach to the AVR and PSS design using domain partitioning, robust pole-replacement, and adaptive control [5–7]. Moreover, recently several control methods have been made to coordinate the various requirements for stabilization and voltage regulation within one new control structure [1,8–12].

The published papers on the AVR and PSS coordination concerning the applied methodologies can be classified into four categories. These four methodology groups are identification/prediction technique, switching concept, intelligent technique, and optimization/robust control approach.

The complexity of the identification process and the resulting model is an important problem of the first category. It will be more difficult for a large-scale power system with a high dimension and time-varying and nonlinearity properties. Here, the control strategies of the first category are briefly discussed and some of their advantages and disadvantages are demonstrated.

A robust control method, namely, the internal model control (IMC), was employed to construct a robust controller for coordinated power system voltage regulator and stabilizer synthesis in Refs [13,14]. The IMC theory states that control can be achieved only if the control system encapsulates some representation of the plant to be controlled. In this method, plant output is predicted by using the plant model. A high performance of IMC is achieved when the open-loop system is stable and the exact model is available. In practice, often the exact model is not available and the open-loop system is also unstable. Therefore, first the system should be stabilized by a feedback controller, and then the prestabilized system must be used for the application of the conventional IMC to obtain the final robust coordinator.

Using the model predictive control (MPC) method, a trade-off between the AVR and PSS performance requirements is provided in Ref. [15]. The method uses the two-axis model of the synchronous generator to perform control action signal. For this purpose, the current and voltage components are used to compute the field voltage, which is required to satisfy the voltage regulation and small signal stability [15]. The MPC method is initially suitable for systems with large time constants, while power system small-signal stability is a small timescale phenomenon. Moreover, time-variance and the high nonlinearity of power systems, as well as the complexity of the controller structure are the other problems of the mentioned control strategy.

The switching concept-based coordination method for transient stability and voltage regulation is addressed in Refs [9,16]. Using the trial and error method, for each specific fault, a unique switching time is obtained. In Ref. [16], direct feedback linearization (DFL) is used to design a robust controller to improve the voltage regulation and transient stability. In the proposed method, the line reactance and infinite bus voltage are assumed as constants. The DFL idea is to algebraically transform nonlinear system equations into the linear ones, so that linear control techniques can be applied. In this work, the postfault voltage and prefault line reactance are considered as uncertain parameters and then the DFL technique is applied to obtain a robust coordinator. The design problem is finally transformed to solve an algebraic Riccati equation (ARE).

The obtained DFL compensating control law through the solution of the ARE can be represented as follows:

$$K_1 = f(x); x = [\Delta\delta \ \omega \ \Delta P_e] \tag{7.1}$$

where δ is the angle, ω is the frequency, P_e is the electrical power, and $f(x)$ is a linear function with respect to $\Delta\delta$, ω, and ΔP_e. By differentiating terminal voltage equation in the linearized power system model, Equation (7.1) can be obtained in the term of x vector. Similarly, for the postfault condition, the voltage control law can be obtained as

$$K_2 = f(y); y = [\Delta v_t \ \omega \ \Delta P_e]. \tag{7.2}$$

Following a fault, the control signal K_1 is employed to keep the generators in synchronism; then in the postfault, the feedback law switches to K_2, which is employed to enhance the power system performance [10,18]. Figure 7.3 shows the block diagram realization of this approach, schematically.

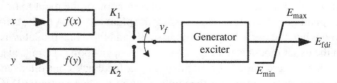

Figure 7.3. Switching-based coordination method.

Several intelligent approaches, specifically fuzzy logic-based voltage regulation and oscillation damping coordinator designs are reported over the years. The methods introduced in Refs [10,11] belong to the third category. The nonrobustness drawback of the switching concept-based method is compensated by weighing the controller output according to the operating points. A global controller to improve transient stability and to achieve satisfactory postfault voltage level in the presence of disturbances is introduced in Ref. [10]. In this work, two trapezoid-shaped membership functions (μ_δ, μ_v) were used such that

$$\mu_\delta(z) = 1 - \mu_v(z) \tag{7.3}$$

where

$$z = \sqrt{\alpha_1 \omega^2 + \alpha_2 (\Delta v_t)^2} \tag{7.4}$$

In the transient period, μ_δ becomes the dominant value, while in the postfault state μ_v becomes the dominant value. According to the operating condition, the control laws are weighted to get a satisfactory performance by performing final coordinating control action signal v_f:

$$v_f = \mu_\delta f(x) + \mu_v f(y) \tag{7.5}$$

where $f(x)$ is the control signal of the DFL controller (7.1), and $f(y)$ is the control signal of voltage regulator (7.2). In fact, the membership functions determine the participation rate of each controller. A block diagram of the mentioned controller is shown in Fig. 7.4. As can be seen, in addition to the complexity of the overall control framework, an extra unit is required to generate the weights.

In Ref. [19], the particle swarm optimization (PSO) technique is used to solve the simultaneous damping and synchronization problem through minimizing of a comprehensive damping index. Some features of the PSO such as less computation time and few memory requirements make it attractive for solving the present optimization problem. Using a linearized dynamic model without considering uncertainty and neglecting the exciter dynamic decreases the effectiveness of the mentioned methodology for real-world power systems. The application of optimal and robust control techniques for coordinating AVR and PSS are given in several papers [1,8,12,20].

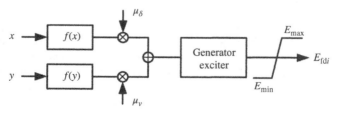

Figure 7.4. Weighed switching-based coordination method.

Complexity of control structure, numerous unknown design parameters, as well as neglecting real constraints can be seen in most of the published coordination control techniques. In practice, usually controllers with a simple structure are preferable. On the other hand, the authors' experience shows that although conventional PSS and AVR systems may be incapable of obtaining desirable dynamic performances for a wide range of operating conditions, the electric power industry still is too conservative to open the conventional control loops and test the new/advanced coordination units, because of probable risks and bugs.

The mentioned challenge is addressed in Refs [1,12] by providing a simple gain vector feedback in parallel with the conventional AVR and PSS units. The design objectives are formulated via a robust static output feedback control problem, and the optimal static gains are obtained using an iterative linear matrix inequalities algorithm.

7.3 A ROBUST SIMULTANEOUS AVR–PSS SYNTHESIS APPROACH

This section presents a methodology to enhance the stability and voltage regulation of existing real power systems without opening their conventional PSS and AVR units. The methodology provides a simple gain vector in parallel with the conventional control devices. The design objectives are formulated via an H_∞ static output feedback (H_∞-SOF) control problem and the optimal static gains are obtained using an iterative linear matrix inequalities (ILMI) algorithm. This work is fully presented in Ref. [1].

The proposed controller includes proportional gains and uses the measurable signals, so it has considerable promise for implementation, especially in a multimachine power system. In fact, the proposed control strategy attempts to make a bridge between simplicity in control structure and robustness in stability and performance to satisfy the AVR and PSS objectives, simultaneously.

To demonstrate the efficiency of the control methodology, some real-time nonlinear laboratory tests have been performed on a four-machine infinite bus system using the Analog Network Simulator (ANS) at the Research Laboratory of the Kyushu Electric Power Company (KEPCO) in Japan. The obtained results are compared with the conventional AVR–PSS design.

7.3.1 Control Framework

The overall control structure using the SOF control design for a given power system is shown in Fig. 7.5, where the PSS and AVR blocks represent the existing conventional power system stabilizer and voltage regulator. Here, the electrical power signal Δp_{ei} is considered as the input signal for the PSS unit. The static feedback controller (optimal gain vector) uses the terminal voltage Δv_{ti}, electrical power Δp_{ei} and machine speed $\Delta \omega_i$ as input signals. The Δv_{refi} and d_i show the reference voltage deviation and system disturbance inputs, respectively.

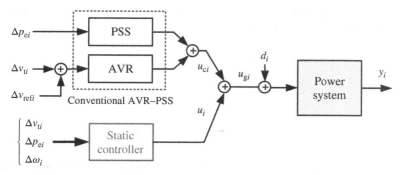

Figure 7.5. Overall control structure.

Consider the linearized model of a given power system "i" in the following state-space form:

$$\dot{x}_i = A_i x_i + B_{1i} w_i + B_{2i} u_i$$
$$z_i = C_{1i} x_i + D_{12i} u_i \qquad (7.6)$$
$$y_i = C_{2i} x_i$$

where x_i is the state variable vector, w_i is the disturbance and area interface vector, u_i is the control input vector, z_i is the controlled output vector, and y_i is the measured output vector. The A_i, B_{1i}, B_{2i}, C_{1i}, C_{2i}, and D_{12i} are real matrices/vectors with appropriate dimensions.

Using standard H_∞-SOF configuration [1] while considering appropriate controlled output signals results in an effective control framework, which is shown in Fig. 7.6. This control structure adapts the H_∞-SOF control technique with the described power system control targets and allows direct trade-off between voltage regulation and closed-loop stability by mere tuning of a gain vector.

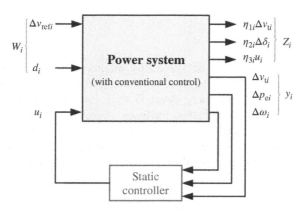

Figure 7.6. The proposed H_∞-SOF control framework.

Here, disturbance input vector w_i, controlled output vector z_i, and measured output vector y_i are considered as follows:

$$w_i^T = \begin{bmatrix} \Delta v_{refi} & d_i \end{bmatrix} \tag{7.7}$$

$$z_i^T = \begin{bmatrix} \eta_{1i}\Delta v_{ti} & \eta_{2i}\Delta\delta_i & \eta_{3i}u_i \end{bmatrix} \tag{7.8}$$

$$y_i^T = \begin{bmatrix} \Delta v_{ti} & \Delta p_{ei} & \Delta\omega_i \end{bmatrix} \tag{7.9}$$

The Δv_{ti} and Δp_{ei} can be easily expressed in terms of system states, and the η_{1i}, η_{2i}, and η_{3i} are constant weights that must be chosen by the designer to get the desired closed-loop performance. Since the vector z_i properly covers all significant controlled signals that must be minimized by an ideal AVR–PSS design, it is expected that the proposed robust controller could be able to satisfy the voltage regulation and stabilizing objectives, simultaneously. Since the solution must be obtained through minimizing of the H_∞ optimization problem, the designed feedback system should satisfy the robust stability and voltage regulation performance for the overall closed-loop system. Moreover, the developed ILMI algorithm (which is described in the next section) provides an effective and flexible tool to find an appropriate solution in the form of a simple static gain controller.

7.3.2 Developed Algorithm

The formulation of static output feedback stabilization generally leads to a bilinear matrix inequalities (BMI) problem, which is a nonconvex problem. As described in Ref. [21], system (A, B, C) is stabilizable via the SOF if and only if there exist $P > 0, X > 0$ (symmetric and positive-definite matrices) and K_i satisfying the following quadratic matrix inequality:

$$\begin{bmatrix} A^TX + XA - PBB^TX - XBB^TP + PBB^TP & (B^TX + K_iC)^T \\ B^TX + K_iC & -I \end{bmatrix} < 0 \tag{7.10}$$

This kind of problem can be solved by an iterative algorithm that may not converge to an optimal solution. Here, in order to solve the H_∞-SOF, an iterative LMI algorithm has been used. The key point is to formulate the H_∞ problem via a generalized static output stabilization feedback such that all eigenvalues of $(A-B K_i C)$ shift toward the left half plane in the complex s-plane, to close to feasibility of (7.10). The general static output feedback control theory [1] gives a family of internally stabilizing SOF gain matrix defined as K_{sof}. Here, the desirable solution K_i is an admissible SOF law

$$u_i = K_iy_i, K_i \in K_{sof} \tag{7.11}$$

such that

$$\|T_{zi\,wi}(s)\|_\infty < \gamma^*, |\gamma - \gamma^*| < \varepsilon \tag{7.12}$$

where ε is a small positive number. The performance index γ^* indicates a lower bound such that the closed-loop system is H_∞ stabilizable. The optimal performance index (γ), can be obtained from the application of a full dynamic H_∞ dynamic output feedback control method. The proposed algorithm, which gives an ILMI solution for the mentioned optimization problem includes the following steps [1]:

Step 1: Set initial values and compute the generalized system $(\overline{A}_i, \overline{B}_i, \overline{C}_i)$ and $\overline{X}$ as given in (7.13), for the given power system including conventional AVR and PSS units.

$$\overline{X} = \begin{bmatrix} X & 0 & 0 \\ 0 & I & 0 \\ 0 & 0 & I \end{bmatrix}, \quad \overline{A}_i = \begin{bmatrix} A_i & B_{1i} & 0 \\ 0 & -\gamma I/2 & 0 \\ C_{1i} & 0 & -\gamma I/2 \end{bmatrix}, \quad \overline{B}_i = \begin{bmatrix} B_{2i} \\ 0 \\ D_{12i} \end{bmatrix}, \quad \overline{C}_i = \begin{bmatrix} C_{2i} & 0 & 0 \end{bmatrix} \tag{7.13}$$

For this purpose, the matrix $\mathbf{A}_i$ has the following form, and the elements of other matrices in (7.13) can be obtained based on the structure of the used excitation system, AVR and PSS units:

$$\mathbf{A}_i = \begin{bmatrix} A_{gi(4\times4)} & 0_{(4\times m)} \\ 0_{(m\times4)} & A_{ci(m\times m)} \end{bmatrix} \tag{7.14}$$

The "c" is used for the conventional AVR–PSS system, A_{gi} is obtained from the linearized state space model of targeted generator, which can be expressed as $\dot{x}_{gi} = A_{gi}x_{gi} + B_{gi}u_{gi}$. A modeling method to obtain a fourth-order linear state-space model for a typical generator is explained in Appendix C.

Step 2: Set $i = 1$, $\Delta\gamma = \Delta\gamma_0$ and let $\gamma_i = \gamma_0 > \gamma$. $\Delta\gamma_0$ and γ_0 are positive real numbers.

Step 3: Select $Q > 0$, and solve $\overline{X}$ from the following ARE:

$$\overline{A}_i^T \overline{X} + \overline{X}\overline{A}_i - \overline{X}\overline{B}_i\overline{B}_i^T\overline{X} + Q = 0 \tag{7.15}$$

Set $P_1 = \overline{X}$.

Step 4: Solve the following optimization problem for $\overline{X}_i$, K_i, and a_i. Minimize a_i subject to the LMI constraints

$$\begin{bmatrix} \overline{A}_i^T\overline{X}_i + \overline{X}_i\overline{A}_i - P_i\overline{B}_i\overline{B}_i^T\overline{X}_i - \overline{X}_i\overline{B}_i\overline{B}_i^T P_i + P_i\overline{B}_i\overline{B}_i^T P_i - a_i\overline{X}_i & \left(\overline{B}_i^T\overline{X}_i + K_i\overline{C}_i\right)^T \\ \overline{B}_i^T\overline{X}_i + K_i\overline{C} & -I \end{bmatrix} < 0 \tag{7.16}$$

$$\overline{X}_i = \overline{X}_i^{\mathrm{T}} > 0. \tag{7.17}$$

Denote a_i^* as the minimized value of a_i.

Step 5: If $a_i^* \leq 0$, go to step 8.

Step 6: For $i > 1$, if $a_{i-1}^* \leq 0$, $K_{i-1} \in K_{\mathrm{sof}}$ is an H_∞ controller and $\gamma^* = \gamma_i + \Delta\gamma$ indicates a lower bound such that the system is H_∞ stabilizable via SOF control. Go to step 10.

Step 7: If $i = 1$, solve the following optimization problem for $\overline{X}_i$ and K_i:
Minimize trace $(\overline{X}_i)$ subject to the LMI constraints (7.16–7.17) with $a_i = a_i^*$.
Denote $\overline{X}_i^*$ as the $\overline{X}_i$ that minimized trace $(\overline{X}_i)$. Go to step 9.

Step 8: Set $\gamma_i = \gamma_i - \Delta\gamma$, $i = i + 1$. Then do steps 3–5.

Step 9: Set $i = i + 1$ and $P_i = \overline{X}_{i-1}^*$, then go to step 4.

Step 10: If the obtained solution (K_{i-1}) satisfies the gain constraint, it is desirable; otherwise, retune constant weights (η_i) and go to step 1.

The vector $\eta_i = \begin{bmatrix} \eta_{1i} & \eta_{2i} & \eta_{3i} \end{bmatrix}$ is a constant weight vector that must be chosen by the designer to get the desired closed-loop performance. The selection of these weights depends on the specified voltage regulation and damping performance objectives. In fact an important issue with regard to the selection of the weights is the degree to which they can guarantee the satisfaction of design performance objectives. It is notable that η_{3i} sets a limit on the allowed control signal to penalize fast changes, large overshoot with a reasonable control gain to meet the feasibility and corresponding physical constraints. Therefore, the selection of constant weights entails a compromise among several performance requirements.

One can simply fix the weights to unity and use the method with regional pole placement technique for performance tuning. Here, for the sake of weight selection, the following steps are simply considered through the proposed ILMI algorithm:

Step 1: Set initial values for $\begin{bmatrix} \eta_{1i} & \eta_{2i} & \eta_{3i} \end{bmatrix}$, for example, [1 1 1].

Step 2: Run the ILMI algorithm.

Step 3: If the ILMI algorithm gives a feasible solution such that it satisfies the robust H_∞ performance and gain constraints, the assigned weights vector is acceptable. Otherwise, retune η_i and go to step 2.

The proposed iterative LMI algorithm shows that if we simply perturb $\overline{A}_i$ to $\overline{A}_i - (a/2)I$ for some $a > 0$, then we will find a solution of the matrix inequality (7.16) for the performed generalized plant. That is, there exist a real number $(a > 0)$ and a matrix $P > 0$ to satisfy inequality (7.16). Consequently, the closed-loop system matrix $\overline{A}_i - \overline{B}_i K \overline{C}_i$ has eigenvalues on the left-hand side of the line $\Re(s) = a$ in the complex s-plane. Based on the idea that all eigenvalues of $\overline{A}_i - \overline{B}_i K \overline{C}_i$ are shifted progressively toward the left half plane through the reduction of a. The given generalized eigenvalue minimization in the proposed iterative LMI algorithm guarantees this progressive reduction.

7.3.3 Real-Time Implementation

To illustrate the effectiveness of the proposed control strategy, a real-time experiment was performed at the ANS laboratory. For the purpose of this study, a longitudinal four-machine infinite bus system is considered as the test system. A single line representation of the study system is shown in Fig. 7.7. Although in the given model the number of generators is reduced to four, it closely represents the dynamic behavior of the West Japan Power System [1]. The most important global and local oscillation modes of the actual system are included. For the study system, the local and global low-frequency oscillation modes are around 1.5 and 0.3 Hz, respectively. Each unit is a thermal unit and has its own conventional excitation control system as shown in Fig. 7.1a (for units 2 and 3) and Fig. 7.8 (for units 1 and 4).

Each unit has a full set of governor–turbine system (governor, steam valve servo-system, high-pressure turbine, intermediate-pressure turbine, and low-pressure turbine), which is shown in Fig. 7.9. The generators, lines, conventional excitation system, and governor–turbine parameters are given in Table C.1 to Table C.5 in Appendix C.

Unit 1 is selected to be equipped with a robust controller, and therefore our objective is to apply the control strategy described in the previous section to the controller design for unit 1. The whole power system has been implemented in the ANS laboratory. Figure 7.10 shows the overview of the applied laboratory experiment devices including the control/monitoring desks. A digital oscilloscope and a notebook computer are used for monitoring purposes.

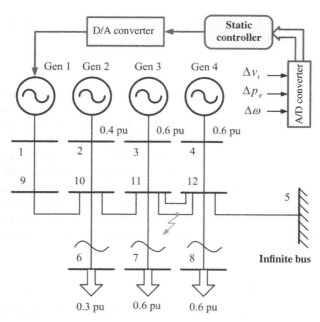

Figure 7.7. Four-machine infinite bus power system.

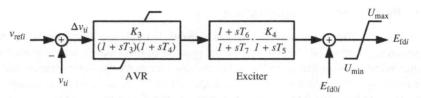

Figure 7.8. Excitation control system for units 1 and 4.

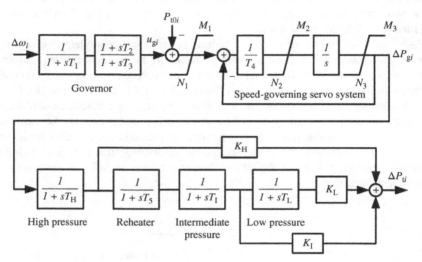

Figure 7.9. Speed governing and turbine system.

The proposed control loop (Fig. 7.11) has been built in a personal computer connected to the power system using a digital signal processing (DSP) board equipped with the analog to digital (A/D) and digital to analog (D/A) converters as the physical interfaces between the personal computer and the ANS hardware. In Fig. 7.11, the input/output scaling blocks are used to match the PC-based controller and the ANS hardware, signally. High-frequency noises are removed by appropriate low-pass filters.

Then, applying the proposed H_∞-SOF control methodology, an optimal gain vector is obtained as $K_{1,SOF} = \begin{bmatrix} 9.5899 & 7.8648 & 1.2990 \end{bmatrix}$. The considered constraints on limiters and control loop gains are set according to the real power system control units and close to ones that exist in the conventional AVR and PSS units. The constant weight vector is obtained as $\eta_1 = \begin{bmatrix} 0.25 & 0.1 & 5 \end{bmatrix}$.

7.3.4 Experiment Results

The performance of the closed-loop system using the proposed optimal gain vector (OGV) in comparison with a pure conventional AVR–PSS system is tested in the

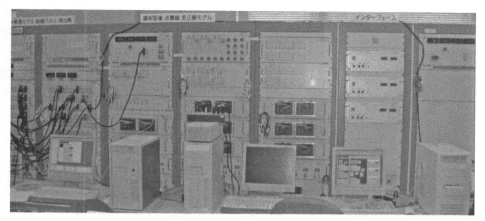

Figure 7.10. Performed laboratory experiment: control/monitoring desks.

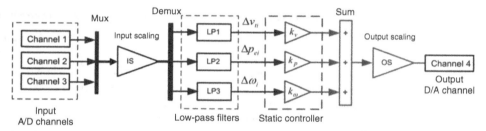

Figure 7.11. The performed computer-based control loop.

presence of voltage deviation, faults, and system disturbance. The configuration of the applied conventional PSS, which was accurately tuned by the system operators, is illustrated in Fig. 7.1b. The conventional PSS parameters are listed in Table C.5 (Appendix C).

During the first test scenario, the output setting of unit 1 is fixed at 0.5 pu. Figure 7.12 shows the electrical power, terminal voltage, and machine speed of unit 1, following a fault on the line between buses 11 and 12 at 2 s. To have a more critical situation, the faulted line is isolated from the network just after four cycles from the fault. It can be seen that the system response is quite improved using the designed feedback gains.

Furthermore, the size of the resulting stable region by the proposed method is significantly enlarged in comparison with the conventional AVR–PSS controller. To show this fact, the critical power output from unit 1 in the presence of a three-phase to ground fault is considered as a good measure. To investigate the critical point, the real power output of unit 1 is increased from 0.3 pu (the setting of the real power output from the other units is fixed at the values shown in Fig. 7.7). Using the conventional AVR–PSS structure, the resulting critical power output from unit 1 will be 0.31 pu [1]; and in the case of tight tuning of conventional control parameters it cannot be higher than

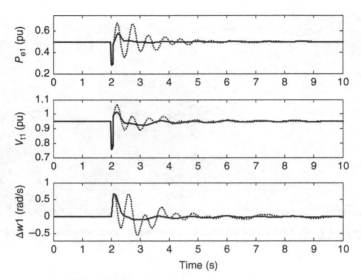

Figure 7.12. System response for a fault between buses 11 and 12. Solid line (robust control); dotted line (conventional control).

0.52 pu. For the proposed control method, the critical power output, as shown in Table 7.1, is increased to 0.94 pu. The system response for a fault between buses 11 and 12, when the output setting of unit 1 is increased to 0.7 pu, is shown in Fig. 7.13.

In the second test case, the performance of the designed controllers was evaluated in the presence of a 0.05 pu step disturbance injected into the voltage reference input of unit 1 at 20 s. Figure 7.14 shows the closed-loop response of the power systems fitted with the conventional control and the proposed robust control design. Better performance is achieved by the developed control strategy. In the next scenario, the closed-loop system response is examined in the face of a step disturbance (d_i) at 20 s. The result is shown in Fig. 7.15. Comparing the experiment results shows that the robust design achieves robustness against the voltage deviation, disturbance, and line fault with a quite good voltage regulation and damping performance.

Finally, to demonstrate the simultaneous damping of local (fast) and global (slow) oscillation modes, filtering analysis has been performed. The laboratory results for the speed deviation of unit 1, following a fault on the line between buses 11 and 12, are shown in Fig. 7.16 (in this experiment, the fault happened at 2 s and the output setting of unit 1 was fixed at 0.45 pu).

TABLE 7.1. Critical Power Output of Unit 1

Control Design	Critical Power Output, pu
Proposed design	0.94
Conventional AVR–PSS	0.52

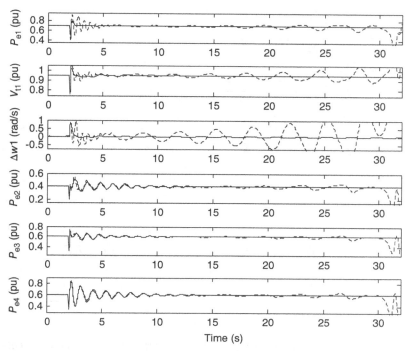

Figure 7.13. System response for a fault between buses 11 and 12, while the output setting of unit 1 is fixed to 0.5 pu. Solid line (robust control); dotted line (conventional control).

7.4 A WIDE-AREA MEASUREMENT-BASED COORDINATION APPROACH

Increased needs for electrical energy as well as environmental concerns and growing attempts to reduce dependency on fossil fuel resources have caused many power system industries to set ambitious targets for renewable power generation. Wind power is recognized as the most important renewable energy source (RES) because of its economic and technical prospects. It is shown that increasing the penetration level could seriously affect power system dynamics [22–24]. Owing to the worldwide focus on connecting a major volume of renewable generation sources to the networks in future years, the coordinated AVR–PSS design problem in the presence of RESs becomes more significant than in the past. However, using simplified assumptions/linear models in the recent proposed control methodologies restricts their applications in the presence of RES dynamics.

This section addresses a tuning algorithm based on the described angle–voltage ($\Delta\delta$–ΔV) graph in Section 4.2. The tuning-based control methodology is used for the coordination of the AVR–PSS system in large-scale power systems concerning wind power penetration. The controller is designed in such a way that all the

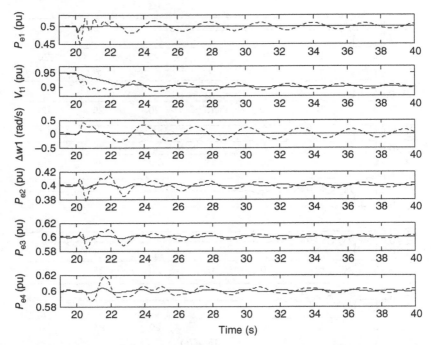

Figure 7.14. System response for a 0.05 pu step change at the voltage reference input of unit 1. Solid line (robust control); dotted line (conventional control).

inherent nonlinearities and uncertainties are considered in the system and hence, the introduced controller is reliable. The proposed controller uses the wide-area measurements data.

The proposed methodology employs the normalized phase difference versus voltage deviation plane (described in Chapter 4) to design a control strategy that is independent of the fault type and the case study. The proposed controller uses wide-area measured voltage and angle signals.

7.4.1 High Penetration of Wind Power

Impacts of renewable energy options especially wind power on the power system dynamics and control are discussed in Section 5.5. References [25] and [26] show that the associated controllers with doubly fed induction generators (DFIGs) separate the mechanical dynamics from the electrical ones, and therefore no electromechanical mode exists. In other words, the DFIGs do not have a detrimental impact on the small-signal stability. Moreover, the transient stability is not of concern considering the DFIGs, and also the fault performance or ride through of the synchronous generator will be even better. Therefore, the DFIGs cannot affect power system dynamics seriously in such a way that it would be an appropriate case study to demonstrate the

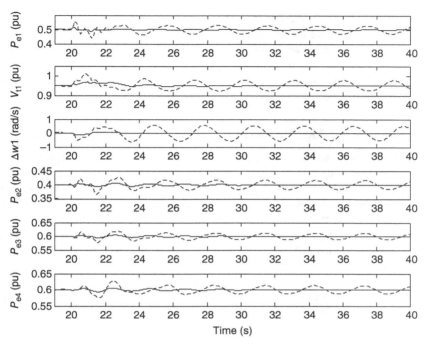

Figure 7.15. System response for a step disturbance at 20 s. Solid line (robust control); dotted line (conventional control).

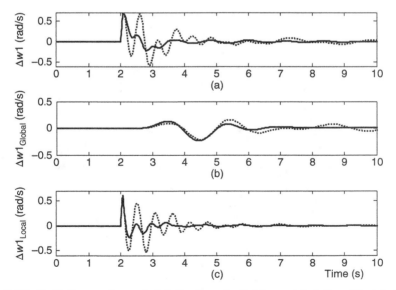

Figure 7.16. Oscillation modes analysis following a fault: (a) speed deviation, (b) global mode, and (c) fast mode. Solid line (robust control); dotted line (conventional control).

ability of the proposed coordination strategy. However, as previously stated the fixed speed induction generators (FSIGs) directly affect power system dynamics by interchanging reactive power with the connected system and hence significantly influence the controller performance. Implementation of auxiliary controllers such as a static VAR compensator (SVC) is mandatory to maintain power system integrity as addressed in this section.

7.4.2 Developed Algorithm [2]

Figure 7.17 demonstrates the generalized version of Fig. 4.3 (given in Chapter 4), which describes the operating region of the AVR, PSS, and SVC. After a description of the required borders to design a coordinated controller, the mechanism on how the variation of the AVR, PSS, and SVC gains could improve power system performance is explained. As already mentioned, the controller gains are retuned in order to conduct each connected generator to the (1, 1) point in the normalized $(\Delta\delta - \Delta V)$ plane. Owing to the detrimental impact of torques in phase with rotor angle deviations on oscillatory stability and also taking into account subsynchronous resonance, the control strategy should be considered in such a way that the AVRs always encounter gains reduction. For a generator with less voltage deviation, variation of AVR gain could enhance the system performance. In other words, if voltage deviation for a generator is smaller than 0.707, decreasing of AVR gain causes an increment of voltage deviation and hence, the generator is conducted to the desired region. The mechanism on which the PSS gain affects the power system dynamics is described next.

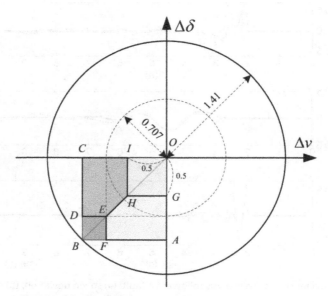

Figure 7.17. Generalized stable region of the power system.

Consider the swing equation as follows:

$$\frac{2H}{\omega_0}\frac{d^2\theta}{dt^2} = P_m - P_e \tag{7.18}$$

where H, ω_0, θ, P_m, P_e are inertia constant, synchronize speed, rotor angle, mechanical power input, and electrical power output, respectively.

The reference for the rotor rotation is considered as a frame that rotates at the synchronized speed. Therefore, θ could be explained by

$$\theta = \omega t - \omega_0 t + \theta_0 \tag{7.19}$$

where θ_0 is the initial position of the rotor. By replacing Equation (7.19) in (7.18), the following equation is obtained based on the speed variations:

$$2H\frac{d(\Delta\omega)}{dt} = P_m - P_e \tag{7.20}$$

It is well known that the PSS produces a torque in phase with positive deviations of speed [1]. For a negative speed deviation, that is, $P_m < P_e$, decreasing of PSS gain accelerates the rotor and increases the rotor angle deviations. Therefore, for a generator with a normalized phase difference smaller than 0.707, the PSS gain should be reduced during the control process. For positive speed deviations, that is, $P_m > P_e$, the PSS gain should be increased in order to conduct the generator with less phase deviations to the desired region. For a generator with both variations smaller than 0.707, a minor variation identifies the required control action to improve system performance. Regarding the above statements, three control laws could be explained as

1. Every generator located in the IHEDC area encounters decreasing of PSS gain in the proposed control strategy.
2. Every generator located in the GAFEH area encounters decreasing of AVR gain in the proposed control strategy.
3. Every generator located in the OGHI area encounters increasing of SVC gain in the proposed control strategy.

The proposed control methodology updates the controller gains by adding a correction term to the old ones in order to improve the system performance. Later, the details on calculating the controller gains are given.

Here, for a located generator in the SVC reaction region, increasing of SVC gain forces the generator operation to the reaction region of the AVR and PSS. Thereafter, the same procedure as steps 1 and 2 enhances the system performance. Therefore, the proposed criterion could be used for the coordinated design of the AVR, PSS, and SVC in multimachine power systems.

For the implementation of the proposed online control strategy, a combination of switching strategy and simple negative feedback in a discrete manner is employed. First, according to the position of each generator in the normalized $\Delta\delta - \Delta v$ plane (Chapter 4), an appropriate control action is selected based on the three control laws presented earlier. Then, the applied feedback tries to conduct generators to the desired region. The implemented feedback employs the angle between phase and voltage deviations (α) as input signal. Note that the set point for the employed feedback is considered as 45°, which is an ideal α at the (1, 1) point. Regarding the time horizon of voltage instability (about 20 s) and transient instability (2 s [27]), the sampling time interval could be selected equal to 2 s. However, the proposed control strategy employs the first sample at a smaller time (less than 2 s) to improve its reliability. The proposed control algorithm can be explained as follows:

Step 1: Set the first sample interval ($T_i = T_{S1}$).

Step 2: Determine the position of each generator in the normalized plane by Δv, $\Delta\delta$, and α.

Step 3: Select a proper control action based on the generators position and control laws.

Step 4: Calculate new control parameters (K_{PSS}, K_{AVR}, K_{SVC}) as follows:

$$K_{pss}^{new} = 1 - \tan|(\alpha - 45)|K_{pss}^{old} \qquad (7.21)$$

$$K_{AVR}^{new} = 1 - \tan|(\alpha - 45)|K_{AVR}^{old} \qquad (7.22)$$

$$K_{svc}^{new} = 1 + \tan|(\alpha - 45)|K_{svc}^{old} \qquad (7.23)$$

Step 5: Set $T = T + \Delta T$ and return to step 2.

where $\Delta T \leq 2$.

Note that, the proposed method is applicable for any load/generation changes. However, as already mentioned for the positive phase deviations, the PSS gain must be increased instead of decreasing. In other words, for the positive phase deviations, that is, $\Delta\delta > 0$, Equation (7.21) will change to

$$K_{pss}^{new} = 1 + \tan|(\alpha - 45)|K_{pss}^{old} \qquad (7.24)$$

The flow chart representation of the proposed algorithm is shown in Fig. 7.18. Here, the K_{PSS}, K_{AVR}, and K_{SVC} are the AVR, PSS, and SVC gains, respectively. In fact, the problem of a coordinated AVR–PSS design is reduced to the design of a simple switching/feedback-based method, which is quite easy for implementation.

7.4.3 An Application Example

The capability of the proposed criterion and control strategy is investigated after application on several large-scale interconnected power systems. Here, this issue is illustrated using the New York/New England system, a 16-machine, 5-area test system, as shown in Fig. 4.4 (Chapter 4). As mentioned, all generators are represented by a two-axis model and equipped with PSS. The applied model for the AVR and PSS is shown in Fig. 4.5. The SVC block diagram is also presented in Fig. 7.19. The efficiency of the proposed control strategy is examined in the presence of wind turbines (WTs). To investigate the impact of WTs on the coordination problem, two wind farms are employed at buses 9 and 15. The implemented wind farms use simple squirrel cage induction generators without any control capability, which operate at a substantially constant speed, normally referred as the FSIGs [2] (Fig. 7.19).

7.4.4 Simulation Results

In Section 4.3, several disturbance scenarios in both generation and demand sides are applied to the considered case study and the importance of the $\Delta\delta$–Δv graph in stability assessment is demonstrated. Here, the effectiveness of the introduced $\Delta\delta$–Δv graph-based tuning approach for coordination of existing control devices (PSS, AVR, and SVC) to enhance oscillation damping and synchronization of the power system with and without wind power penetration are examined.

7.4.4.1 Case A: Without Wind Power. In the performed simulations without wind power penetration, the first sample is used at 1.5 s ($T_{S1} = 1.5$ s). First, the behavior of the power system after the outage of generator 12 is investigated. Note that without applying the proposed control strategy, as shown in Figs. 4.10 and 4.11, the system encounters transient instability, that is, first swing instability. The system response for this disturbance when using the explained tuning-based coordination strategy is shown in Fig. 7.20. The results demonstrate the efficiency of the proposed criterion and control strategy. Figure 7.20a shows the performance of the power system after applying the proposed control strategy within 4 s for the sake of comparison with the first swing instability without applying the strategy. It can be seen that the proposed control strategy aids the system to keep its transient stability. Figure 7.20b explains the behavior of the system within 30 s in the $\Delta\delta$–Δv plane. It is clear that after a number of control reactions, the system conducts to a stable mode at the steady state. Distribution of the voltage–angle indexes in the normalized $\Delta\delta$–Δv plane is also shown in Fig. 7.20. It can be seen that generators 14–16 are located in the PSS reaction region at the first control action.

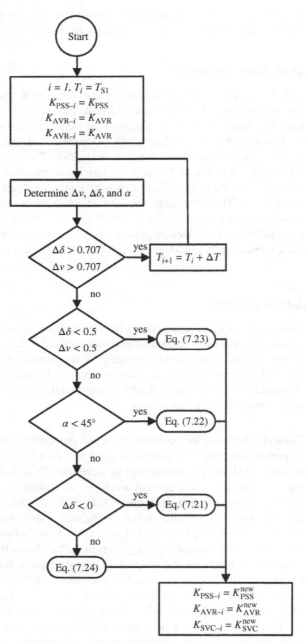

Figure 7.18. Flowchart representation of the control strategy.

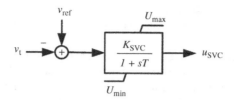

Figure 7.19. The SVC block diagram.

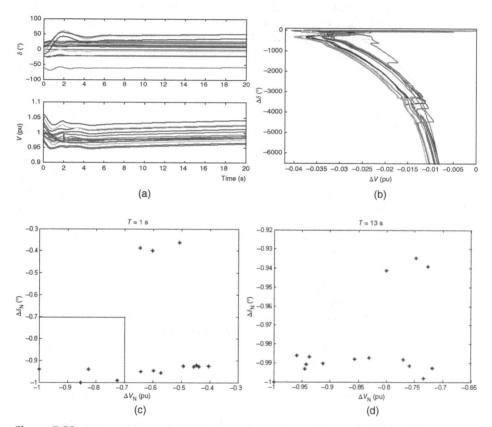

Figure 7.20. System response for the outage of generator 12 in the presence of the proposed control strategy: (a) time-domain results, (b) $\Delta\delta - \Delta V$ graphs, (c) distribution of the normalized parameters in the $\Delta\delta - \Delta V$ plane at 1 s, and (d) at 13 s.

Therefore, decreasing the PSS gains of these generators besides decreasing the AVR gains for some others makes the system stable. It can be seen that the system response is quite improved using the proposed coordination methodology (Fig. 7.20d).

The simulation results show the satisfactory performance of the control strategy and success of the proposed criterion against the load changes as well as generation changes.

As another example, Fig. 7.21 shows the dynamics of the system after outage of generator 16. It can be seen that the system experiences disorders in the oscillatory, angle, and transient stability behavior during simulation time. The applied control strategy eliminates the voltage/rotor angle oscillations and conducts the system to a stable mode in the steady state. Note that without applying the proposed control strategy the system encounters instability. Distribution of the voltage–angle indexes related to the outage of generator 16 in the normalized $(\Delta\delta - \Delta V)$ plane is also shown in Fig. 7.21. At $t = 1.5$ s (Fig. 7.21c), except for two generators, the others are out of the desired region. According to the assigned control laws, by decreasing the PSS gains for this set of generators at the first control action, the system tries to keep its integrity. It can be seen that all the generators have been located in the desired region at the steady state (Fig. 7.21d).

7.4.4.2 Case B: With Wind Power. Case B is performed to emphasize the efficiency of the proposed method for the power system in the presence of wind farms. In the performed simulations concerning wind power penetration, the T_{S1} will be fixed at 1.8 s. Figure 7.22 shows the dynamic response of the power system following outage of G12, concerning 8% wind power penetration. It can be seen that by the penetration of wind power, generators 14–16 try to be isolated from the system and operate in islanding mode. Therefore, the system is going into an unstable mode. The nonsimilar phase characteristics of generators besides the diverged terminals voltage behavior in Fig. 7.22b also exhibit instability of the system. Distribution of the voltage–angle indexes in the normalized $(\Delta\delta - \Delta V)$ plane is shown in Fig. 7.22c.

It can be seen that generators 14–16 are located in the operating region of the SVC. Therefore, the SVC should be implemented in the areas related to these generators. However, because of fewer deviations for generator 16 and taking into account the economic reasons, it seems that the SVC in this region could return the system to a stable mode. For this purpose, at the beginning of simulation, the SVC is considered in the model and after that, increasing of SVC gain improves the system performance.

The dynamic behavior of the power system after applying the proposed control strategy is shown in Fig. 7.23. The proposed control strategy conducts the system to a stable mode within several tuning steps. The initial and final values (controller gains at steady state) of control parameters are given in Table 7.2. Distribution of the normalized parameters in the proposed criterion for outage of G12 in the presence of wind farms and SVC can be seen in Fig. 7.23. From Fig. 7.23c, it is quite clear that all the generators are moved to the desired region at steady state.

7.4.4.3 Discussion. In the performed simulations, it was seen that in the transient period (based on the system inertia), interactions between the AVR and PSS gains improve the system performance and thereafter, reduction of AVR gains

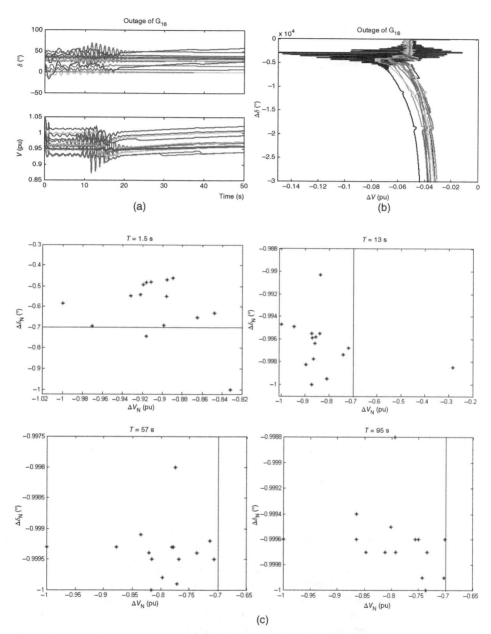

Figure 7.21. System response for the outage of generator 16 by applying the proposed control strategy: (a) time-domain results, (b) $\Delta\delta-\Delta V$ graphs, (c) distribution of the normalized parameters in the $\Delta\delta-\Delta V$ plane at 1.5, 13, 57, and 95 s.

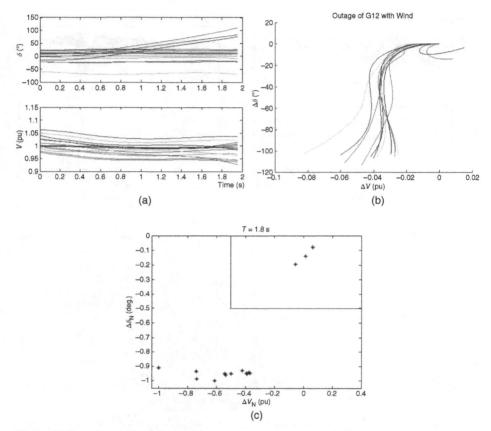

Figure 7.22. System response (with conventional control system) for the outage of generator 12 in the presence of wind power: (a) time-domain results, (b) $\Delta\delta-\Delta V$ graphs, and (c) distribution of the normalized parameters in the $\Delta\delta-\Delta V$ plane at 1.8 s.

conducts the system to a stable mode. In other words, the PSSs contribute to maintain the transient stability. On the other hand, to avoid oscillatory instability, the torques in phase with rotor angle deviations, that is, AVRs gains, always decrease. The contribution of the PSS for keeping transient stability is discussed well by Kundur [27].

A practical discontinues excitation control (DEC), which employs switching strategy, was used at the Ontario hydro power plant to achieve stability and voltage regulation. The DEC addresses transient stability excitation control (TSEC) to improve stability. The TSEC is switched off from the system after 2 s (transient period) to avoid oscillatory instability. In other words, the DEC employs a fixed time-based switching strategy. However, the transient period can be varied from 2 s to about 10 s based on the system size and the mode status. Therefore, the fixed time-based switching strategy could affect the system performance.

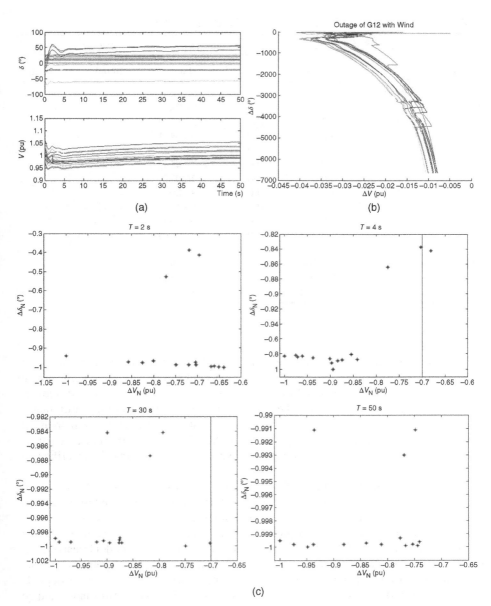

Figure 7.23. System response for the outage of generator 12 following application of the proposed method in the presence of wind power: (a) time-domain results, (b) $\Delta\delta - \Delta V$ graphs, and (c) distribution of the normalized parameters in the $\Delta\delta - \Delta V$ plane at 2, 4, 30, and 50 s.

TABLE 7.2. The Initial and Final Control Gains for Outage of G12

Generator Number	PSS Gains (K_{PSS})		AVR Gains (K_{AVR})		SVC Gains (K_{SVC})	
	Initial	Final	Initial	Final	Initial	Final
1	100	100	100	100.000	–	–
2	100	100	100	81.7805	–	–
3	100	100	100	80.2319	–	–
4	100	100	100	76.6451	–	–
5	100	100	100	75.1585	–	–
6	100	100	100	74.4325	–	–
7	100	100	100	76.0875	–	–
8	100	100	100	82.0945	–	–
9	100	100	100	80.1594	–	–
10	100	100	100	100.000	–	–
11	50	50	100	82.2249	–	–
13	110	110	100	81.7274	–	–
14	100	75.8369	100	64.7924	–	–
15	100	79.9705	100	64.8628	–	–
16	100	82.3612	100	81.5471	50	58.3413

The proposed algorithm in this section fulfills this concern by using an adaptive angle-based switching strategy. Based on the system operation, the synchronizing and damping torques are changed to achieve a satisfactory performance. Moreover, decreasing the AVR gains in the proposed control methodology eliminates probable subsynchronous resonance (SSR) in the system. It is shown that increasing the AVR gains could encounter a system with SSR [28]. For the present case study, the SSR phenomenon, which results from increasing AVR gains, is described in Ref. [2].

Due to the different kinds of uncertainty and parameter perturbation, load variations, and modeling errors, considerable efforts have been devoted to the design of robust excitation control. A conic programming method, which moves eigenvalues corresponding to the unstable and poorly damped modes to the left-hand side of the s-plane is introduced in Ref. [25]. In general, successfulness of all model linearization-based control methodologies severely depends on the relevant operating points. Considering various operating points, that is, the robustness issue, could degrade the controller performance by increasing the size of evaluation, control laws, and so on. However, the addressed graphical criteria in Chapter 4 and control strategy in the present section is absolutely independent of the operating point and hence, the robustness feature completely and successfully can be met. Moreover, the proposed control strategy relies on the basic power system equations and therefore it is applicable to any power system without any query. Simulation results prove the satisfactory performance of the control strategy under contingencies, as well as in normal conditions.

Simplicity of analysis using the developed graphical tool and model independency of the proposed coordinated control strategy can be considered as the main advantages of the present work.

7.5 INTELLIGENT AVR AND PSS COORDINATION DESIGN

As mentioned before, following a disturbance in a power system, while a high-gain fast-response AVR improves the large-signal transient stability, it also has a detrimental effect on the oscillation stability and has a converse effect on the PSS operation for the transient stability. Therefore, in a fault situation it is necessary to have coordination between the two controllers.

This section presents the performance of intelligent fuzzy-based coordinated control for the AVR and PSS to prevent losing synchronism after a major sudden fault and to achieve appropriate postfault voltage level in multimachine power systems. The AVR and PSS gains are adaptively adjusted to guarantee the power system stability after faults.

To change AVR and PSS gains, one or more generators in each area must be equipped with the proposed fuzzy logic coordinator. The fuzzy logic unit accepts normalized deviations of terminal voltage and phase difference (described in Sections 4.2 and 7.4) as inputs and tunes AVR and PSS gains.

7.5.1 Fuzzy Logic-Based Coordination System

Figure 7.24a shows the overall scheme for the proposed fuzzy logic-based coordinator. The normalization and scaling blocks are not fuzzy variables and have been designed to improve the coordination performance. As shown, the fuzzy unit contains four major components: fuzzification, fuzzy rule-base, inference mechanism, and defuzzification.

Here, terminal voltage deviation and angle deviations are considered as input signals of the fuzzy system. These input signals are selected to guarantee the synchronism and stability enhancement of the power system. Furthermore, to limit the input signal deviations, a normalization method is applied. For this purpose, a method, which is conceptually shown in Fig. 7.24b, is used for both voltage and angle deviations. The $\Delta\delta_{i-N}$ (ΔV_{i-N}) demonstrates the normalized phase difference (voltage deviation). The $\Delta\delta_{i-N}$ and ΔV_{i-N} can vary in the range of $[-1, 1]$ for all generators.

Fuzzification plays an important role in dealing with uncertain information, which might be objective or subjective in nature. The fuzzification block represents the process of making crisp quantity into fuzzy variables. In fact, the fuzzifier converts the crisp input to a linguistic variable using the assigned membership functions in the fuzzy knowledge base. Fuzziness in a fuzzy set is characterized by the membership functions. Using appropriate membership functions, the input/output quantities are converted to desirable linguistic variables, which finally specify the quality of control commands.

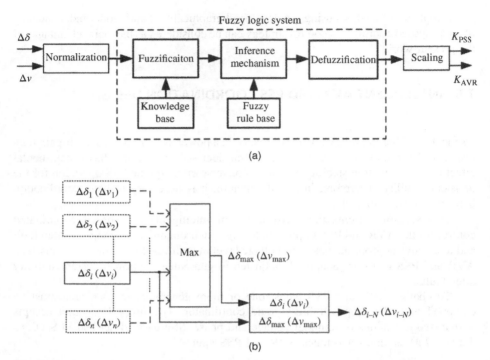

Figure 7.24. (a) Fuzzy-logic-based coordinator. (b) Normalization method for the input signals.

Here, the membership functions corresponding to the input variables are arranged as negative large (NL), negative small (NS), zero (ZR), positive small (PS) and positive large (PL), and for the output variables they are arranged as very small (VS), small (S), medium (M), and large (L). The number of membership functions for each variable may affect the quality of the output control command. In the present work, five membership functions are defined for the input signals and four membership functions are defined for output signals. The membership functions of input and output variables (for the case study shown in Fig. 6.8) are demonstrated in Fig. 7.25.

The knowledge rule base consists of information for linguistic variables definitions (database) and fuzzy rules (control commands). The concepts associated with a database are used to characterize fuzzy control rules and a fuzzy data manipulation in the fuzzy logic controller. A lookup table is made based on discrete universes that define the output of a controller for all possible combinations of the input signals. A fuzzy system is characterized by a set of linguistic statements in the form of "if–then" rules. Fuzzy conditional statements make the rules or the rule set of the fuzzy controller. Finally, the inference engine uses the if–then rules to convert the fuzzy input to the fuzzy output.

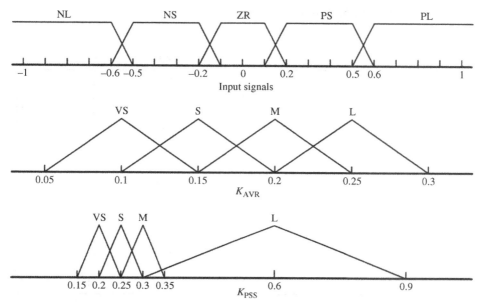

Figure 7.25. Membership functions for input signals and output signals (K_{AVR}, K_{PSS}).

Fuzzy inference is the kernel of a fuzzy logic system. With two inputs and five linguistic terms, 25 rules were developed for each output, which is given in Table 7.3 [17]. The antecedent parts of each rule are composed by using the AND function. Here, the Mamdani fuzzy inference system is employed. The centroid method is used for defuzzification.

7.5.2 Simulation Results

The efficiency of the fuzzy logic-based coordinated control for AVR–PSS is tested on the 2-area, 4-machine, 11-bus test system, which is shown in Fig. 6.8. The detailed bus data, line data, and dynamic characteristics for the machines, exciters, and loads are given in Ref. [27]. All generators use AVR and PSS controllers with the following transfer functions:

$$G_{AVR}(s) = \frac{200}{0.01s + 1} \qquad (7.25)$$

$$G_{PSS}(s) = \frac{300s(0.05s + 1)(3s + 1)}{(10s + 1)(0.02s + 1)(5.4s + 1)} \qquad (7.26)$$

The power system has two areas with two generators in each area. The nominal system is operating with 413 MW being sent from area 1 to area 2. The AVR–PSS units of

TABLE 7.3. Fuzzy Rule Base for (a) PSS Gain and (b) AVR Gain

(a)		ΔV_N				
K_{PSS}		NL	NS	ZR	PS	PL
$\Delta \delta_N$	NL	VS	VS	VS	VS	S
	NS	S	S	S	M	M
	ZR	S	S	M	M	M
	PS	S	S	M	M	M
	PL	M	L	L	L	L

(b)		ΔV_N				
K_{AVR}		NL	NS	ZR	PS	PL
$\Delta \delta_N$	NL	VS	S	VS	VS	S
	NS	VS	S	S	VS	L
	ZR	VS	S	S	S	L
	PS	VS	S	S	S	M
	PL	S	M	M	M	L

generator 2 in area 1 (connected to the swing bus in the system) and generator 4 at area 2 (with a high power rate) are considered to be coordinated/controlled by the designed fuzzy system.

Since, the time horizon for transient instability is almost about 3–5 s and this time horizon for small-signal instability is about 10–20 s, the fuzzy control unit will be activated in the system following 1 s after faults. As the first test scenario, the power system performance is investigated after outage of generator 3; and at the second test scenario, 1000 MW load has been suddenly added to the power system [17].

Outage of generator 3 can be accounted for as a severe fault in the present power system. Figure 7.26 shows the time domain responses of the generator outputs following a fault without the fuzzy coordinator system. Here, the P_a is accelerator power, V_t is terminal voltage and w is generator speed. Strong oscillations in accelerator powers are shown. It can be also seen that the power system loses its synchronism, and two areas are separated. Therefore, generator 4 cannot be able to supply whole the load in area 2 and voltage is collapsed in this area.

Figure 7.27 shows the system response (voltage, speed, and accelerator power) for the same fault in the presence of using a fuzzy logic system for tuning the AVR–PSS units in generators 2 and 4. The fuzzy systems maintain the synchronism and return the terminal voltages to the acceptable levels.

In the second test scenario, 1000 MW load step disturbance is applied to the system. The system responses without and with the proposed fuzzy logic-based coordination system are shown in Figs. 7.28 and 7.29. It is shown that using conventional AVR–PSS

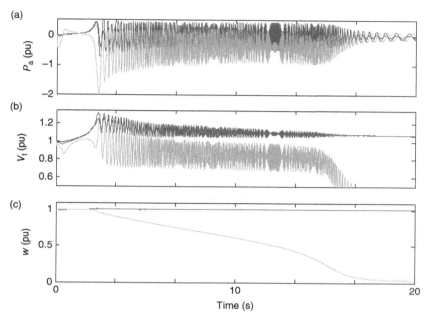

Figure 7.26. System generator response following outage of generator 4 without intelligent coordinator: (a) accelerator power, (b) terminal voltage, and (c) speed.

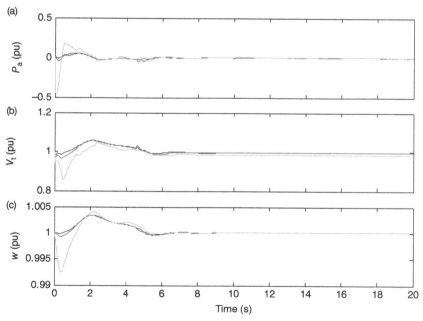

Figure 7.27. System generator response following outage of generator 4 with intelligent coordinator: (a) accelerator power, (b) terminal voltage, and (c) speed.

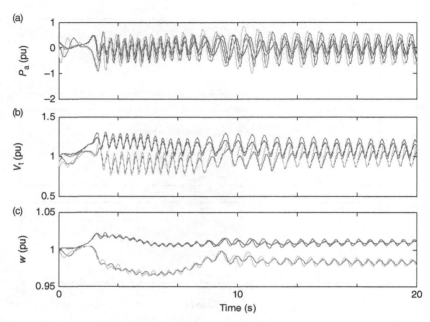

Figure 7.28. System generator response following 1000 MW load step increase without intelligent coordinator: (a) accelerator power, (b) terminal voltage, and (c) speed.

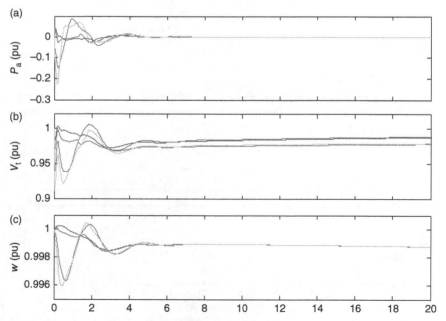

Figure 7.29. System generator response following 1000 MW load step increase with intelligent coordinator: (a) accelerator power, (b) terminal voltage, and (c) speed.

control, the system falls to an oscillatory instability and the synchronism is also lost. On the other hand, with the fuzzy logic-based coordination, the synchronism remains and the oscillations are successfully removed.

7.6 SUMMARY

Power system stability and voltage regulation have been considered as important control issues for secure system operation over many years. Currently, because of expanding physical setups and penetration of RESs, the mentioned problems become more significant than in the past. Moreover, because of conflict behavior between stability and voltage regulation the successful achievement of both goals using independent control strategies is difficult.

In order to achieve stability and voltage regulation simultaneously, three control strategies using robust H_∞ control, $\Delta\delta-\Delta v$ plot, and intelligent fuzzy system are addressed. The first control strategy is developed based on the H_∞-SOF control technique via an iterative LMI algorithm. The proposed method was applied to a four-machine infinite bus power system, through the ANS laboratory real-time experiment, and the results are compared with a conventional AVR–PSS design. The performance of the resulting closed-loop system is shown to be satisfactory over a wide range of operating conditions. As shown in the nonlinear real-time simulation results, the proposed coordination through a new optimal feedback loop has brought a significant effect on improving the power system performance and widening the stable region.

The second control strategy uses a criterion in the normalized phase difference versus the voltage deviation $\Delta\delta-\Delta v$ plane. Based on the introduced criterion, an adaptive angle-based switching strategy and negative feedback are combined to obtain a robust control methodology against load/generation disturbances. The proposed method is applied to the IEEE 68-bus test system in the presence of wind power penetration. The obtained results based on the proposed control strategy exhibit satisfactory performance over a wide range of operating conditions.

As the third control scheme, a fuzzy logic-based coordination system for optimal tuning of AVR–PSS gains is introduced. The fuzzy system uses the normalized terminal voltage and phase angle deviations as input signals and generates the appropriate AVR and PSS gains. The closed-loop system performance is examined on the IEEE 11-bus test system.

The resulting controllers are not only robust but also allow direct effective trade-off between voltage regulation and damping performance. Furthermore, because of simplicity of structure, decentralized property, ease of formulation, and flexibility of design methodology, they are practically desirable.

REFERENCES

1. H. Bevrani, T. Hiyama, and Y. Mitani, Power system dynamic stability and voltage regulation enhancement using an optimal gain vector, *Control Eng. Pract.*, **16**, 1109–1119, 2008.

2. H. Golpira, H. Bevrani, and A. H. Naghshbandi, An approach for coordinated automatic voltage regulator power system stabiliser design in large-scale interconnected power systems considering wind power penetration, *IET Gen. Transm. Distrib.*, **6**(1), 39–49, 2012.

3. F. P. Demello and C. Concordia, Concepts of synchronous machine stability as affected by excitation control, *IEEE Trans. Power Apparatus Syst.*, **88**, 316–329, 1969.

4. G. Rogres, *Power System Oscillations*, Kluwer Academic Publishers, London, 2000.

5. V. A. Venikov and V. A. Stroev, Power system stability as affected by automatic control of generators: some methods of analysis and synthesis, *IEEE Trans. Power Apparatus Syst.*, **90**, 2483–2487, 1971.

6. H. M. Soliman and M. M. F. Sakar, Wide-range power system pole placer, *Proc. Inst. Electr. Eng. C*, **135**(3), 195–200, 1988.

7. O. P. Malik, G. S. Hope, Y. M. Gorski, V. A. Uskakov, and A. L. Rackevich, Experimental studies on adaptive microprocessor stabilizers for synchronous generators. In: *IFAC Symposium on Power Systems and Power Plant Control, Beijing, China*, 1986, pp. 125–130.

8. A. Heniche, H. Bourles, and M. P. Houry, A desensitized controller for voltage regulation of power systems, *IEEE Trans. Power Syst.*, **10**(3), 1461–1466, 1995.

9. Y. Wang and D. J. Hill, Robust nonlinear coordinated control of power systems, *Automatica*, **32**(4), 611–618, 1996.

10. Y. Guo, D. J. Hill, and Y. Wang, Global transient stability and voltage regulation for power systems, *IEEE Trans. Power Syst.*, **16**(4), 678–688, 2001.

11. N. Yadaiah, A. G. D. Kumar, and J. L. Bhattacharya, Fuzzy based coordinated controller for power system stability and voltage regulation, *Electr. Power Syst. Res.*, **69**, 169–177, 2004.

12. H. Bevrani and T. Hiyama, Stability and voltage regulation enhancement using an optimal gain vector. In: *Proceedings of the IEEE PES General Meeting, Canada*, 2006.

13. K. T. Law, D. J. Hill, and N. R. Godfrey, Robust controller structure for coordinated power system voltage regulator and stabilizer design, *IEEE Trans. Control Syst. Technol.*, **2**, 220–232, 1994.

14. K. T. Law, D. J. Hill, and N. R. Godfrey, Robust co-ordinated AVR-PSS design, *IEEE Trans. Power Syst.*, **9**, 1218–1225, 1994.

15. M. Saidy, A unified approach to voltage regulator and power system stabilizer design based on predictive control in analogue form, *Electr. Power Energy Syst.*, **19**, 103–109, 1997.

16. Y. Wang, D. J. Hill, R. H. Middleton, and L. Gao, Transient stability enhancement and voltage regulation of power systems, *IEEE Trans. Power Syst.*, **8**, 620–627, 1993.

17. R. Khezri and H. Bevrani, Fuzzy-based coordinated control design for AVR and PSS in multimachine power systems, 13th Iranian Conf. on Fuzzy Systems (IFCS), Qazvin, Iran, Aug. 2013.

18. C. Zhu, R. Zhou, and Y. Wang, A new nonlinear voltage controller for power systems, *Int. J. Electr. Power*, **19**, 19–27, 1997.

19. A. M. El-Zonkoly, Optimal tuning of power systems stabilizers and AVR gains using particle swarm optimization, *Expert Syst. Appl.*, **31**, 551–557, 2006.

20. H. Boules, S. Peres, T. Margotin, and M. P. Houry, Analysis and design of a robust coordinated AVR/PSS, *IEEE Trans. Power Syst.*, **13**, 568–575, 1998.

21. Y. Y. Cao, J. Lam, Y. X. Sun, and W. J. Mao, Static output feedback stabilization: an ILMI approach, *Automatica*, **34**(12), 1641–1645, 1998.

22. D. Gautam, V. Vittal, and T. Harbour, Impact of increased penetration of DFIG-based wind turbine generators on transient and small signal stability of power system, *IEEE Trans. Power Syst.*, **24**, 1426–1434, 2009.

23. H. Bevrani, A. Ghosh, and G. Ledwich, Renewable energy sources and frequency regulation: survey and new perspectives, *IET Renew. Power Gen.*, **4**(5), 438–457, 2010.

24. H. Bevrani and A. G. Tikdari, An ANN-based power system emergency control scheme in the presence of high wind power penetration. In: L. F. Wang, C. Singh, and A. Kusiak, editors, *Wind Power Systems: Applications of Computational Intelligence*, Springer Book Series on Green Energy and Technology, Springer, Heidelberg, 2010, pp. 215–254.

25. R. A. Jabr, B. C. Pal, and N. Martins, A sequential conic programming approach for the coordinated and robust design of power system stabilizers, *IEEE Trans. Power Syst.*, **25**, 1627–1637, 2010.

26. B. C. Pal and F. Mei, Modelling adequacy of the doubly fed induction generator for small-signal stability studies in power systems, *IET Renew. Power Gen.*, **2**, 181–190, 2008.

27. P. Kundur, *Power System Stability and Control*, McGraw-Hill, New York, 1994.

28. G. Rogers, *Power System Oscillations*, Kluwer Academic Press, Boston, MA, 1999.

8

WIDE-AREA MEASUREMENT-BASED EMERGENCY CONTROL

To prevent power system blackout following a severe contingency, an emergency control action may be needed. Serious load generation imbalance, which is usually the result of a severe contingency, may lead the system to cascading failures and even blackout. Load shedding (LS) is a well-known emergency control scheme used to curtail the loads that could not be supplied in an acceptable time duration, before loss of the remaining power. A general review on the role of disturbance size/location, shed load block size/location, and shed delay time in the effectiveness of LS actions shows that a wide-area LS approach is expected to be a useful solution candidate for developing LS schemes to offer better coordination (considering cascading failures).

Frequency and voltage are more frequent decision indices in the emergency control strategies. Most LS schemes proposed so far separately use voltage and frequency information via underfrequency and undervoltage LS (UFLS/UVLS) schemes. Furthermore, the underfrequency and undervoltage relays usually work in the power system without any coordination. In this chapter, the necessity of considering both voltage and frequency indices to achieve an effective and comprehensive LS strategy is emphasized. It is also clarified that this problem will be more dominant for a wide-area power system with renewable energy sources (RESs) such as wind power turbines.

Power System Monitoring and Control, First Edition. Hassan Bevrani, Masayuki Watanabe, and Yasunori Mitani.
© 2014 John Wiley & Sons, Inc. Published 2014 by John Wiley & Sons, Inc.

In the present chapter, after a background on the LS problem, an LS scheme using both voltage and frequency information with the described graphical tool in Section 4.4 (Chapter 4) is introduced. Then the application of electromechanical wave propagation for the emergency control issue in a wide-area power system is emphasized.

8.1 CONVENTIONAL LOAD SHEDDING AND NEW CHALLENGES

Load shedding can be considered as an effective emergency control scheme in power systems. The LS can be started in the form of underfrequency or undervoltage LS schemes. The UFLS and UVLS work based on a significant drop in frequency and voltage, respectively. For example, following a severe event, when the total load in the system is higher than the available generation capacity, the frequency will go down. In this case, some loads may be shed to bring the frequency back within the permitted limit.

8.1.1 Load Shedding: Concept and Review

The LS is an emergency control action to ensure system stability by curtailing system load. The emergency LS would only be used if the frequency/voltage falls below a specified frequency/voltage threshold. Typically, the LS protects the system against excessive frequency or voltage decline by attempting to balance real and reactive power supply and demand in the system.

If the power system is unable to supply its active and reactive load demands, the underfrequency and undervoltage conditions will be intense. To prevent the postload shedding problems and overloading, the location bus for the LS will be determined based on the load importance, cost, and distance to the contingency location.

The number of LS steps, amount of load that should be shed in each step, the delay between the stages, and the location of shed load are the important issues that should be determined in an LS algorithm. An LS scheme is usually composed of several stages. Each stage is characterized by frequency/voltage threshold, amount of load, and delay before tripping. The objective of an effective LS scheme is to curtail a minimum amount of load, and provide a quick, smooth, and safe transition of the system from an emergency situation to a normal equilibrium state.

Most common LS schemes are the UFLS schemes, which involve shedding predetermined amounts of load if the frequency drops below specified frequency thresholds. The UVLS schemes, in a similar manner, are used to protect against excessive voltage decline. A comprehensive and useful guideline for UFLS strategies can be found in Ref. [1]. The UFLS schemes typically curtail a predetermined amount of load at specific frequency thresholds. In comparison with the UVLS, the UFLS time delays are small; they are about 0.2 s.

There are many published research reports on the UVLS schemes. A guideline for the UVLS can be found in Ref. [2]. In Ref. [3] some reasons about using UVLS beside UFLS are given. In this article, the transmission line outage is introduced as a case in which voltage may drop before observing excessive depression in system frequency. In this situation, frequency may drop slowly and voltage may collapse before UFLS relays can sense the contingency situation.

It is noteworthy that the UVLS schemes are commonly used to avoid voltage collapse problems. Furthermore, the dip voltage transients are local problems where some practical methods such as the use of special protective relays can be used to avoid voltage collapses following a local voltage drop. These protective relays should cover the function standards to sense voltage magnitude and to trip the breaker if a dip voltage drop is observed. The UVLS schemes should consider such breaker-interlock protective functions by coordinating their time delay and thresholds.

The time delays in UVLS schemes should be set above 3 s to avoid false tripping [2]. In Ref. [4], it is mentioned that the UVLS time delays must not be too small to mitigate the reliability criteria introduced by the National Electrical Reliability Coordinating Council (NERCC). Furthermore, Ref. [4] proposes a slope permissive UVLS design to solve the problem by shedding load before this long time delay following prediction of a considerable voltage drop.

To represent the dynamics of emergency control actions, the incremented/decremented step behavior is usually used. For instance, for a fixed UFLS scheme, the function of LS in the time domain could be considered as a sum of the incremental step functions of $\Delta P_j u(t - t_j)$, as given in Equation (5.8).

There are various types of UFLS/UVLS schemes discussed in literature and applied by the electric utilities around the world. A classification divides the existing schemes into *static* and *dynamic* (or *fixed* and *adaptive*) LS types. Static LS curtails the constant block of load at each stage, while dynamic LS curtails a dynamic amount of load by taking into account the magnitude of disturbance and dynamic characteristics of the system at each stage. Although the dynamic LS schemes are more flexible and have several advantages, most real-world LS plans are of static type. The improved UFLS/UVLS algorithms are usually adaptive. The adaptive schemes usually offer larger amount of load at the first LS step following large disturbances [5]. The frequency thresholds can also be biased to using disturbance magnitude to shed load at higher frequency levels in dangerous contingencies [5–9]. The rate of frequency change is a frequent additional parameter for estimating the disturbance magnitude and adjusting the frequency thresholds [6–12].

There are two basic paradigms for LS [5,9]: a *shared* LS paradigm and a *targeted* LS paradigm. The first paradigm appears in the well-known UFLS schemes, and the second paradigm in some recently proposed wide-area LS approaches. Using simulations for a multiarea power system, it is easy to illustrate the difference between these two paradigms, following generation loss in one area.

Sharing LS responsibilities (such as that induced by the UFLS) is not necessarily an undesirable feature and can be justified on a number of grounds. For example, shared LS schemes tend to improve the security of the interconnected regions by allowing generation reserve to be shared. Furthermore, the LS approaches can be indirectly used to preferentially shed the least important load in the system. However, sharing the LS can have a significant impact on interregion power flows and, in certain situations, might increase the risk of cascade failure.

Although both shared and targeted LS schemes may be able to stabilize overall system frequency/voltage, the shared LS response leads to a situation requiring more power transmission requirements. In some situations, this increased power flow might cause line overloading and increase the risk of cascade failure.

8.1.2 Some Key Issues

As the use of RES increases worldwide, there is also a rising interest in their impacts on power system operation and control. The important impacts of a large penetration of variable power generators in the area of operation and control can be summarized in the directions of regional overloading of transmission lines in normal operation as well as in emergency conditions, reduction of available tie-line capacities due to the large load flows, frequency performance, grid congestions, increasing need for balance power and reserve capacity, increasing power system losses, increasing reactive power compensation, and impact on the system security and economic issues [6].

The distributed power fluctuation (due to the use of variable generations) negatively contributes to the power imbalance, frequency, and voltage deviations. Significant disturbance can cause under/over frequency/voltage relaying and disconnect some lines, loads, and generators. Under unfavorable conditions, this may result in a cascading failure and system collapse. There are few reports on the role of distributed RESs in emergency conditions.

Among all RESs, wind power has attracted more attention over the last three decades. Increasing the penetration of wind turbine generators (WTGs) in a power system may affect its security/stability limits, frequency, voltage, and dynamic behavior [13–18]. The WTGs commonly use the induction generators (IG) to convert wind energy into electrical energy [16,17]. The induction generators can be considered as reactive power consumers. Therefore, the voltage of the system (as well as UVLS) would be affected in the presence of wind turbines, especially in the case of fixed-speed-type WTGs. In Ref. [17], the effects of the doubly fed induction generator (DFIG) and IG type of WTGs on the voltage transient behavior are explained and the disadvantages of the IG type are shown. Some notes are also given in Chapter 5. The loadability of various types of WTGs is compared in Refs [13,14,18], and it is shown that the DFIG has larger loadability than IGs. Frequency nadir in the presence of different types of the WTGs has been compared in Ref. [16]. As argued in the mentioned references, wind turbines affect frequency behavior (and in result UFLS scheme) because they add the amount of inertia in the power system. Both stator and rotor windings of IG-type WTGs are connected directly with the power grid, but in the DFIG type only the stator is directly connected and the rotor is connected through a power electronic-type converter. The IG-type WTGs in turn add much inertia than the DFIG in the power system; and in conclusion, the IG-type WTG's frequency response is better than systems with the DFIG type, with the same conditions.

Some reports have also addressed the impacts of various WTG technologies on the voltage deviation following a contingency event, and have analyzed their influences on the transient voltage stability. It is illustrated that the $P–V$ curve is significantly affected by changing the network topology. Some parameters such as power system reserve and inertia constant are influenced by interconnecting the WTGs on the power system. Therefore, the frequency deviation will be also affected in the presence of the wind turbine [19]. Following a disturbance, the frequency decline and initial rate of frequency change in the presence of IGs is smaller than the DFIGs case. Because of their structure, the IGs add more inertial response to the power system than DFIGs.

Recent works indicate that using frequency gradient for the power system emergency control in the presence of wind turbines needs to be revised. Since in the presence of WTGs, the undesirable oscillations are added to the frequency deviation, the measuring of frequency gradient introduces another difficulty to achieve this variable in emergency control strategies. This issue encourages power design engineers to use $\Delta f / \Delta t$ instead of df/dt [14].

Furthermore, as it is discussed in Section 5.5, as well as in the next section, the voltage and frequency behavior does not address the same results about contingency conditions. This phenomenon encourages researchers to reevaluate the emergency control schemes for the future of the power systems that are integrated with a high penetration of wind power.

Most LS schemes proposed so far separately use voltage and frequency information via the UFLS and UVLS schemes. A majority of published works on the UFLS schemes only consider the active part of load while, by considering the reactive power part, frequency decline is also affected [19]. Furthermore, in the actual power system, the loads contain both active and reactive parts. It is noteworthy that coupling between the active and reactive parts of load ($P–Q$ coupling) can significantly affect the LS schemes. Recently, some research works have been conducted on the necessity of considering both voltage and frequency indices to achieve an effective and comprehensive LS strategy [14].

Another important issue following a severe disturbance in a power system is the coordination between the amount of available spinning reserve and the required emergency control action. Coordination between the amount of spinning reserve allocation and the LS can reduce total costs that generation companies should pay in the emergency conditions.

8.2   NEED FOR MONITORING BOTH VOLTAGE AND FREQUENCY

For the sake of dynamic simulations and to describe/examine the proposed methodology, an updated version of the IEEE nine-bus power system is considered as the test system. A single line diagram for the test system is shown in Fig. 5.13 (Chapter 5). Simulation data and system parameters are given in Ref. [20] and rewritten in Appendix B.

To study the impacts of different types of wind turbines on the voltage and frequency behavior, a large generation loss disturbance in the nine-bus test system is simulated [19]. Generator G2, which is the largest, is tripped at $t = 10$ s for the following cases: without wind turbine, with 10% DFIG-type penetration, with 10% IG-type penetration, and with 10% IG-type wind turbine compensated with a static compensator (STATCOM). Figures 8.1 and 8.2 show the voltage and frequency response following this disturbance,

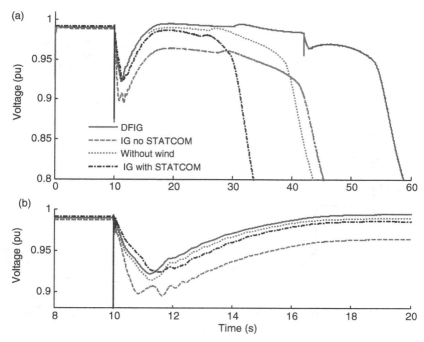

Figure 8.1. (a) Voltage deviation at bus 5 following loss of G2. (b) Zoomed view around 10 s.

respectively. A part of these figures have been already shown in Fig. 5.14. The voltage is monitored at bus 5.

The rate of frequency change is also illustrated in Fig. 8.2. All four cases are unstable. Therefore, an LS scheme is needed to be applied at the first few seconds (following the disturbance) to protect the system from blackouts. The considerable difference between the four cases is demonstrated through the simulation results. Some noteworthy observations can be summarized as follows:

1. In the case of IG-type WTG without STATCOM, the initial voltage drop is the largest (Fig. 8.1), but the initial rate of frequency change is the best (Fig. 8.2).
2. In the case of IG-type WTG compensated with STATCOM, the initial voltage drop is not the worst (Fig. 8.1), but the frequency response is the worst (Fig. 8.2). In this case, the collapsing time is also the shortest. In comparison with the case of IG without the STATCOM, whereas the initial voltage drop is better, the initial rate of frequency change is worse.
3. When DFIG type is used, the time to collapse is longest, initial voltage drop is relatively small, and the frequency response is the best; however, the initial rate of frequency change is large.

From these observations it can be concluded that voltage and frequency response may behave in opposite directions. The voltage and frequency indices cannot monitor the

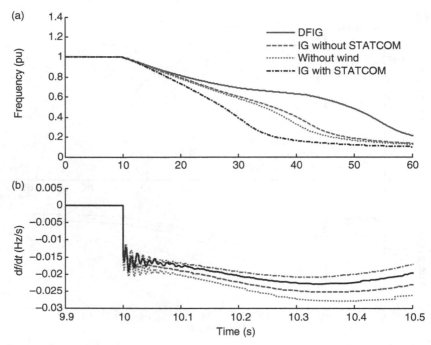

Figure 8.2. Frequency deviation and the rate of frequency change following G2 outage with and without WTGs (DFIG, IG, and IG equipped with STATCOM).

system conditions, especially in the presence of WTGs. The reactive power support devices that should be included with WTG-connected power grids may affect the collapsing time, initial rate of frequency change, and system frequency response. This fact shows that LS schemes that separately use voltage and frequency data need a revision.

Here, an example is given to demonstrate the effectiveness rate of different LS scenarios based on system voltage and/or frequency response. Figure 8.3 shows the voltage and frequency deviations for two different LS scenarios following the same contingency. In these tests, G1 is tripped at 10 s. In scenario 1, only 9% of total system active power is curtailed, while in scenario 2, in addition to 9% active power, 9% of total reactive power is also discarded. Both scenarios shed the load when the frequency falls below 59.7 Hz as used in some existing LS standards such as the Florida Reliability Coordinating Council (FRCC) standard [21]. Considering the frequency and voltage behavior in the two scenarios, some important points are achieved.

A majority of published research works on the UFLS schemes only consider the active part of the load, while by considering the reactive power part, frequency change to be affected as shown in Fig. 8.3b. Furthermore, in the actual power system, the loads contain both active and reactive parts. It is noteworthy that P–Q coupling (coupling between active and reactive parts of the load) can significantly affect the LS schemes. As

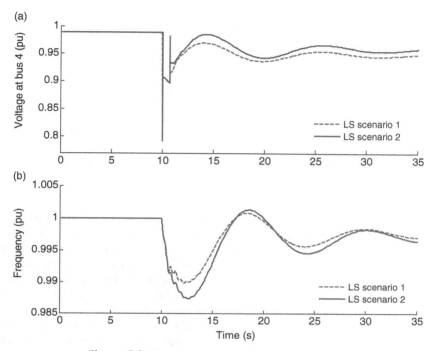

Figure 8.3. System response with two LS scenarios.

argued in Ref. [21], the first-order dynamic model of induction motors is not reliable for the LS because of its inability to capture *P–Q* coupling in the whole operating points.

Figure 8.3 does not show which LS scenario is more effective. The voltage plot illustrates a better performance for scenario 2, while the frequency plot shows an inverse result. These simulations show that by individual monitoring/using of frequency and voltage, there is no guaranty of achieving an effective LS strategy.

8.3 SIMULTANEOUS VOLTAGE AND FREQUENCY-BASED LS

8.3.1 Proposed LS Scheme

As mentioned, the UFLS and UVLS schemes work in the power system without any coordination. The interaction between voltage and frequency responses is neglected in these schemes. Therefore, the risk of LS schemes malfunction can appear in real-world power systems. From previous descriptions and the performed simulation results, it is realized that considering just frequency or voltage indices cannot lead to an effective/optimal LS plan, especially when the reactive power is incorporated in the studies.

TABLE 8.1. The LS Parameters

LS Steps	a_i	b_i	LS, %
A	0.3/60	0.05	9
B	0.6/60	0.12	7
C	0.9/60	0.15	7
D	1.2/60	0.18	6
E	1.5/60	0.2	5
F	1.8/60	0.21	7
$\vdots$	$\vdots$	$\vdots$	$\vdots$

Since the coordination between conventional UFLS and UVLS as separate algorithms is difficult and even impossible, both voltage and frequency should be used in the same LS program simultaneously. With this target in mind, one can define a new graphical analysis tool to monitor both voltage and frequency in one LS algorithm. As shown in Chapter 4 (Section 4.4), the frequency–voltage Δv–Δf graph can be used as a useful graphical tool to see the state of the system following a contingency.

Here, for calculation of the LSCS$_i$ (4.21), the a_i and b_i parameters are considered as given in Table 8.1. It can be seen that in order to present a reasonable comparison between the UFLS scheme and the proposed underfrequency/voltage LS (UFVLS) scheme, the LS parameters are selected as close as possible to the LS strategy introduced by the FRCC [21]. The maximum number of LS steps and the amount of needed loads for shedding per step are also considered similar to the FRCC control plan. Furthermore, the a_i parameters are selected the same as the frequency thresholds used in the FRCC scheme. Therefore, by fixing variable y^t in (4.21) at zero, the proposed algorithm acts similar to the FRCC's UFLS program.

Each step is determined by an ellipse and when the phase trajectory reaches each ellipse, the corresponding LS step will be triggered. Figure 8.4 shows the result of the applied LS steps on the Δv–Δf trajectory plane in the case of G2 outage. The time delay parameter between steps should be considered in the LS algorithm. The voltage and frequency may need to pass through a low-pass filter before entering the algorithm. Existing practical constraints should be also considered in the proposed scheme.

The loads are selected to be triggered at the zones where their LSCS parameter (4.21) leads to an LS step triggering. Actually, when a disturbance happens, the voltage and frequency of a zone will change faster than others; therefore, the loads in this zone will find the highest priority for the LS.

Fine adjustment of delay is significant for achieving a soft frequency/voltage improvement scenario to avoid overshedding and false shedding problems. The UFLS and UVLS schemes are considerably different in time delay magnitudes. The UVLS time delays should be set at about 2–3 s to avoid false shedding. However, typical UFLS time delays are about 0.2–0.3 s [13,14,21]. In order to establish a comprehensive LS scheme, the introduced Δv–Δf plane is divided into four regions, as illustrated in Fig. 8.4.

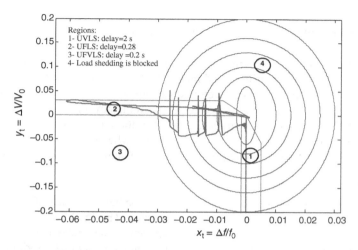

Figure 8.4. The UFVLS scheme considering time delay and permitted shedding regions following loss of G2.

This partitioning lets us define different time delays for different conditions. Region 1 introduces an area with significant drop of voltage and small frequency change. If the Δv–Δf trajectory falls in this region, similar to the UVLS plans, the time delay should be long enough to avoid false shedding. Here, it is fixed at 2 s. On the other hand, region 2 introduces a situation with a high frequency drop and small voltage changes. Time delay in this condition should be also long; however, to prevent LS schemes from falling in a trap made by voltage support devices, it should not be considered too long. Here, similar to the UFLS plans, the time delay is fixed at 0.28 s. The upper and lower borders of this region and the associated time delay may also depend on the magnitude of available reactive power supports, wind power penetration, and wind turbine types. Region 3 introduces an area with significant drop in both voltage and frequency. Therefore, the time delay in curtailing of loads should be reduced. However, to avoid overshedding, it is not permitted to be too small. Here, the time delay for region 3 is fixed at 0.2 s. The LS is blocked in the remaining region (region 4). This region may demonstrate a good condition for the load restoration problem.

8.3.2 Implementation

It is clear that the conventional underfrequency and undervoltage relays cannot support the mentioned graphical tool (elliptical curves). Therefore, it is important to clarify how the proposed methodology can be implemented. The existing practical LS schemes in real power systems mostly use the following technologies [1,2]:

– Breaker interlock LS
– Underfrequency relay LS

– Undervoltage relay LS
– Programmable logic controller (PLC)-based LS

The first method, which is the simplest practical LS method, uses the trip signals of the special generators/tie-lines to directly shed off some load blocks. The second and third methods use undervoltage/underfrequency relays to shed specified load blocks when the locally measured voltage or frequency drops down under some predefined threshold. The most important defect of these three methods is that they do not consider the overall power system circumstance when they produce a command to trip a load. Therefore, they cannot present an optimal and effective LS response for all situations.

The PLC devices have opened a new window to realize new power system control solutions such as the proposed LS in the present chapter. The idea of using PLC is a practical solution for more complicated LS schemes. The PLC-based LS scheme is a more flexible LS method, which uses underfrequency/voltage relay commands as input signals, and then decides to trip some loads based on its predefined priority load levels. The PLC devices can use frequency and voltage signals as analog input (AI) variables to perform the LSCS parameter from (4.21). The result should be evaluated to determine the trigger time of the LS step.

As described beforehand, the proposed UFVLS program uses system frequency and voltage values of different buses. Because the voltage drop is a local problem, the source of the voltage value that would be used in the UFVLS program is not yet determined. It should be updated based on the real-time measured voltages and the location of maximum voltage drop. Figure 8.5 illustrates the flowchart diagram of the proposed UFVLS scheme.

As shown, the selected bus contains the maximum voltage drop. Based on the frequency and selected voltage value, the $LSCS_i$ is calculated. If the $LSCS_i$ is larger than 1, for a proper time period, a predetermined amount of load should be shed from buses near the selected one. This period of time is known as time delay, which is determined based on the real-time Δv–Δf trajectory. The forbidden regions and different time delays are considered as described in the previous subsection.

In the flowchart diagram, the flag parameter is a temporary variable, which used to control the set/reset of timers to adjust the time delays. The timers start if the $LSCS_i > 1$, and stop when they reach a predetermined value (timer (i) > delay (i)). The timer should immediately be reset if $LSCS_i < 1$. The i and i_{max} are the number and maximum number of the LS steps, respectively. It is noteworthy that the illustrated flowchart is completely implementable by standard PLC's programming languages.

8.3.3 Case Studies and Simulation Results

The 9-bus and 24-bus test systems are simulated in the MATLAB/SimPower environment. In both systems, the proposed LS scheme is simulated in two cases: first, it is examined when no WTG is connected to the power grid; second, the ability of the algorithm is examined in the presence of WTGs by another example in each test system.

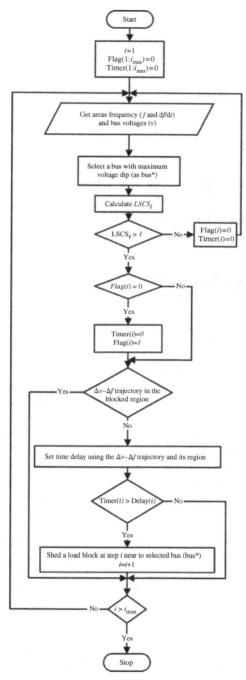

Figure 8.5. UFVLS flowchart.

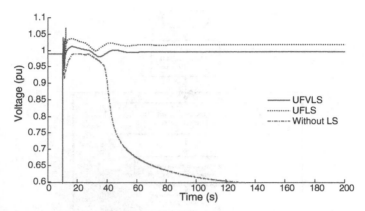

Figure 8.6. Voltage at bus 4 following loss of G2 and using of different LS schemes in nine-bus test system.

8.3.3.1   Nine-Bus Test System. The described LS scheme in the previous section is applied to the nine-bus power grid as a test system, following loss of G2 at $t = 10$ s. The proposed LS scheme is used as an emergency control plan. For the sake of comparison, the conventional UFLS, developed by FRCC [21], is also examined. As shown in Fig. 8.6, both LS methodologies are able to save system stability but the proposed LS presents a better performance.

Overshedding can be seen as a problem of the UFLS scheme. It causes the voltage at bus 4 to stay above its normal level. As shown, in the proposed LS scheme the problem of overshedding is initially removed. Figure 8.4 shows the system state trajectory on the Δv–Δf plane during the LS process. Simultaneous use of voltage and frequency data and good adjustment of the region's time delays are the reasons for this improvement. Figures 8.4 and 8.6 illustrate the response of the system without WTG penetration. However, the proposed scheme is validated on a nine-bus power grid in the presence of high-penetration DFIG-type WTGs with a variable wind speed as another example; the time domain system frequency and voltage response for the new LS scheme are shown in Fig. 8.7.

8.3.3.2   24-Bus Test System. A single line diagram of the 24-bus reliability test system (RTS) is illustrated in Fig. 8.8. The RTS with its full data is introduced in Refs [22,23], and the generators data are selected the same as the given typical data in Ref. [20]. Here, the test system is divided into three areas. While most of the generation is located in area 1 and most of the load is located in area 3, in area 2, the amount of load and generation are approximately equal. Area 1 delivers its overgeneration into areas 2 and 3 through three tie-lines: line 16–19, line 16–14, and line 24–3. All generators use governors with 0.05 pu droop value (R); and also for the generators wherein their nominal output power is larger than 50 MW, a PSS is considered. The load model is a three-phase parallel RLC load, which implements a three-phase balanced load as a parallel combination of the RLC elements. At the specified frequency, the load exhibits

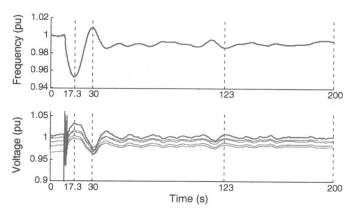

Figure 8.7. System response following G2 outage and proposed LS scheme; test is examined on a nine-bus test system in the presence of WTGs with a variable wind speed.

constant impedance. The active and reactive powers absorbed by the load are proportional to the square of the applied voltage. Four wind farms are connected with the system to introduce a high wind power penetration power system. They supply about 10% of total system demand. All of the wind turbines are of DFIG type.

To validate the proposed LS scheme, several contingencies have been applied to the 24-bus test system. It is investigated that some contingencies affect frequency response more considerably than voltage behavior (a large generator loss near a voltage support device, that is, synchronous condenser at bus 14, like G15 outage), some contingencies affect voltage more considerably than frequency (a far generator like G7), and finally some contingencies significantly affect both voltage and frequency (G23 outage). Contrary to the conventional LS plans, the proposed scheme is expected to stabilize the system in all the mentioned scenarios by curtailing a minimum amount of load.

To have an obvious simulation result, first, the test system without including WTGs is considered. Then, an example will validate it in the presence of WTGs. To demonstrate the capability of the proposed scheme, a large disturbance is applied; outage of lines 16–19 and 16–14 at $t = 2$ s. Following this large disturbance, line 3–24 will be encountered with a high overloading problem. This overloading is larger than its angle stability limit. Therefore, the angle of G1 and G15, which are approximately located at two sides of the line, are separated, and the angle instability phenomena will immediately happen. To save the system in this circumstance, one solution is the use of islanding control.

Islanding control is practically implementable by controlling the interarea coupling devices. If an area observes an external large disturbance from an interconnected area, the islanding control function commands to decouple the connection with that faulted area immediately. These functions can be implemented using hardwired interlocks, breakers, and some special protective relays.

Therefore, the remaining tie-line is tripped and saving the resulting two systems is attempted separately. The main concern is on areas 2 and 3, which encounter an excess load condition. The LS schedule should be used in these areas. This is a large disturbance

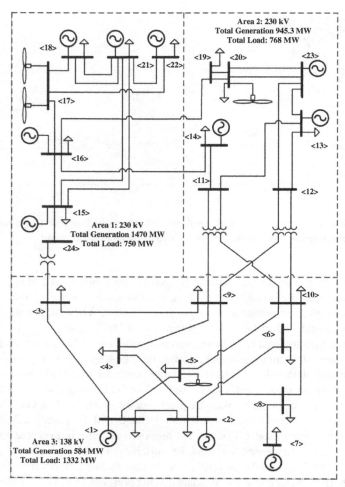

Figure 8.8. Three-area 24-bus test system.

and the simulation results show that application of the conventional LS schemes is not able to restore the system stability. Therefore, an improved version of the previous methods should be used to have a reasonable validation.

As mentioned, most of the recent published works on the emergency control techniques suggest an LS methodology that sheds a larger amount of load with shorter time delay in dangerous contingencies. Here, these points are used to examine the proposed method and to compare it with other works. Therefore, the LS schemes examined here for the mentioned contingency shed a larger amount of load at the first LS step (9% + 7% instead of 9%). The first frequency threshold is also modified ($f_t = 59.8$ Hz instead of 59.7 Hz for the UFLS scheme and $a_1 = 0.02/60$ instead of 0.03/60 for the proposed scheme).

On the other hand, as in the emergency control plans, time is a vital parameter; there is a critical value for the time of emergency control actions to restore the system conditions. With this point in mind, the following scenarios are tested here following the mentioned contingency:

1. The islanding control is done with a 0.4 s time delay at $T_{ic} = 2.4$ s (T_{ic} is the time of islanding control);
2. $T_{ic} = 2.6$ s;
3. $T_{ic} = 2.7$ s; in the 24-bus test system, there is more than one generator at some buses. All of those that have sizes larger than 30 MW are equipped with the PSS in the mentioned cases (cases 1–3). However, in order to have a more critical situation, case 4 is designed.
4. In this case, only one of the generators that is connected to the same bus is equipped with the PSS, and also the T_{ic} is assumed to be 2.4 s.

For the sake of comparison, the UFLS and the proposed LS schemes are used here when the system conditions are completely similar in both LS scenarios with the same shed load amount per step. The simulation results are illustrated in Fig. 8.9. For short islanding control time delays ($T_{ic} = 2.4$–2.6 s), both LS plans succeed in maintaining system stability; however, the proposed scheme presents a more soft post contingency frequency behavior. The advantages of the proposed scheme will be more clarified in the case of worse situations. Increasing time delay and/or decreasing the number of PSS forces the system condition into a critical situation where the UFLS scheme cannot stabilize it, but the proposed scheme is still successful. For larger islanding control time delays ($T_{ic} \geq 2.8$ s), none of the schemes are able to protect the system.

To clarify why the (conventional) UFLS plans work worse than the proposed LS scheme, the voltage and frequency response of the fourth case are illustrated in Fig. 8.10. As shown in this figure, in the first few seconds following the contingency and before loss of line 3–24, voltage drops quickly, while frequency declines slowly. The UVLS time delays do not allow the related (UVLS) relays to cut the loads quickly. Furthermore, the UFLS relays cannot immediately sense the frequency drop. Therefore, although it is a very large disturbance, the LS actions cannot be run on time. It may be dangerous when the time delay of emergency actions is long and the system falls into a critical situation. On the other hand, the voltage is temporarily recovered following manual tripping of line 3–24, but frequency drop is intensified (this shows another demonstrative example to justify the existing conflict between voltage and frequency behaviors). The system frequency quickly drops and it leads the UFLS relays to trip more loads. This is another reason that causes the conventional UFLS schemes to be unsuccessful. Here, it is assumed that there are no load restoration requirements in the system.

Figures 8.11 and 8.12 are presented to answer why the proposed LS is more effective. The elliptic lines in Fig. 8.11 depict the LS thresholds. These lines cross the left-hand side of the x-axis at some points, which are actually the UFLS thresholds. The

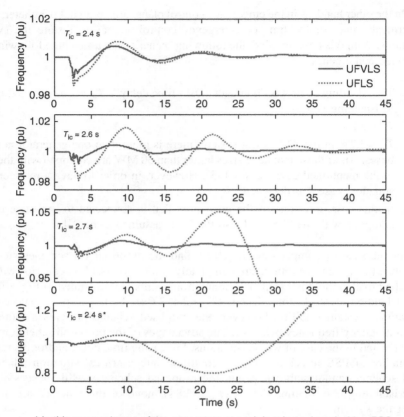

* In this case only one of the generators at each bus is equipped with a PSS.

Figure 8.9. System response using the proposed scheme following loss of line 16-19 and 16-14; line 3-24 is also manually tripped.

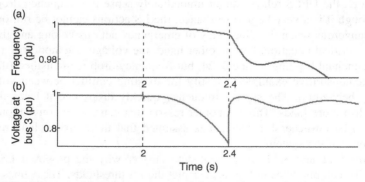

Figure 8.10. Voltage and frequency response following the examined contingency (zoomed view around 2 s).

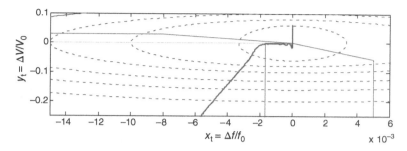

Figure 8.11. The Δv–Δf plot in the first few seconds following the contingency.

Δv–Δf trajectory following the mentioned contingency is also illustrated in the figure. Its map on the left-hand side of the x-axis is the frequency trajectory used in the UFLS, as a conventional LS scheme. It is shown in Fig. 8.11 that the proposed LS thresholds are reached much faster than the UFLS thresholds. If both voltage and frequency parameters decline, the new LS also present a shorter time delay. It is obvious that earlier LS action following a dangerous disturbance, with a large load/generation imbalance, presents a better performance because it blocks voltage/frequency depression. On the other hand, if frequency drops while the voltage stays at a higher level, the LS action is blocked (Fig. 8.4). All of these points help the proposed LS to shed load at an appropriate time and to avoid overshedding, as shown in Fig. 8.12. This figure compares the LS steps versus time for the UFLS and the proposed LS schemes, in case 4.

The contingency and emergency control actions on the 24-bus test system, which have been described earlier, are examined here, but in the presence of high-penetration DFIG-type WTGs (about 10%) with a variable wind speed (line 16–19 and 16–14 are loosed at $t = 10$ s, and also, as an islanding control action, line 3–24 is manually tripped at $t = 10.4$ s). Voltage and frequency behavior in this situation are illustrated in Fig. 8.13. The figure shows the ability of the proposed method to restore the system voltage and frequency responses.

Furthermore, as shown, the variable wind speed makes voltage/frequency have oscillations. The magnitude of oscillations can be amplified by increasing the WTG

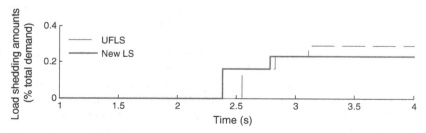

Figure 8.12. Two scenarios of LS steps.

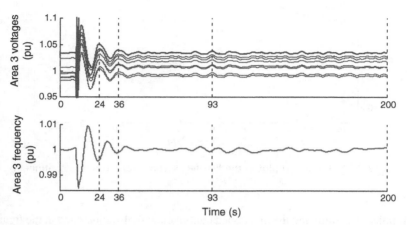

Figure 8.13. System responses following loss of line 16-19 and 16-14; line 3-24 is also manually tripped and the proposed LS scheme is used; test is examined on the 24-bus system in the presence of WTGs with variable wind speed.

penetration. They may deceive LS actions to shed unnecessary amount of load when a frequency/voltage drop emanating from falling of wind speed takes place exactly at the duration of frequency/voltage drop derived from a contingency. Figure 8.13 (like Fig. 8.7) shows that the voltage/frequency behaviors that are affected by variation on wind speed and the existence of the contingency are varied approximately in two opposite directions; voltage improves while frequency is becoming worse and vice versa. It shows that simultaneous consideration of both voltage and frequency parameters for a load shedding scheme makes it powerful to prevent the overshedding problem in the presence of a high penetration of variable speed WTGs.

8.3.4 An Approach for Optimal UFVLS

In the optimal LS programs, the LS thresholds and the amount of LS in each step should be adjusted in order to minimize an objective function. In the conventional UFLS algorithms, the thresholds to be fixed at some frequencies and the objective function will be considered as a function of system frequency behavior and total curtailed loads [24,25].

Based on what has been illustrated in this section, using only frequency data (thresholds and frequency behavior) for finding an optimal LS solution does not lead to the global optimal results. Improving the frequency behavior may disprove the voltage behavior (see the simulation results for the nine-bus test system). Even when improving frequency behavior following large disturbances, just using frequency data may not be effective enough because of voltage behavior that prevents frequency response to drop quickly (see the simulation results for the 24-bus test system).

The proposed UFVLS algorithm, using both voltage and frequency data, makes it possible to adjust the LS parameters optimally. Here, the variables of an optimal UFVLS program that should be adjusted are a_i, b_i, and ΔP_i. The objective function, which demonstrates simultaneous frequency and voltage behavior, can be considered as follows:

$$F_{obj}(Z) = C_1|\widehat{\Delta f}_i| + C_2|\widehat{\Delta v}_i| + C_3 \sum |\Delta P_i| \qquad (8.1)$$

where, $F_{obj}(Z)$ is the objective function, ΔP_i is the LS amount in step i, Z consists of adjustable variables (a_i, b_i, and ΔP_i). $\widehat{\Delta f}$ and $\widehat{\Delta v}$ are the maximum drops in frequency and voltage response, respectively. The variables a_i, b_i, ΔP_i, $\widehat{\Delta f}_i$, and $\widehat{\Delta v}_i$ are bounded variables. Furthermore, C_1–C_3 are constant weights that should be determined by the designer to achieve a desirable LS performance.

8.3.5 Discussion

8.3.5.1 Remarks on Simulation Results.
To clarify why the proposed LS scheme works better than the (conventional) UFLS plans, voltage and frequency response of the fourth scenario of the 24-bus case study is illustrated in Figs. 8.10–8.12. As shown in Fig. 8.10, at the first few seconds following the contingency and before loss of line 3–24, voltage drops quickly, while frequency declines slowly. The UVLS time delays do not allow the related (UVLS) relays to cut the loads quickly; also, although the frequency response is slow, the UFLS thresholds do not arrive on time. Furthermore, the frequency drops quickly following islanding. This causes UFLS actions to overshed, while the UFVLS does not shed more loads because of voltage improvement following the islanding. Figure 8.12 illustrates the difference between the USLS actions and the UFVLS actions. Considering this figure and Fig. 8.11, which shows the Δv–Δf trajectory and the UFVLS thresholds, it is illustrated that the UFVLS scheme sheds loads faster than the UFLS plan. Also, the UFVLS prohibits the overshedding (two steps in the UFVLS and three steps in the UFLS). Figure 8.11 shows that the UFVLS thresholds are reached faster than the UFLS thresholds (cut points of the UFVLS thresholds on the x-axis are the corresponding UFLS thresholds). This prevents frequency to drop more.

8.3.5.2 Need for Further Research.
The present work may provide the first step in considering both voltage and frequency parameters in performing an effective LS scheme, and hence addresses the first stage of promising analysis/synthesis tools on the corrective emergency control actions. More and more research activities are needed to enhance the presented control idea in the future.

Using convex spaces presents a simple LS function, and elliptical curves can simply consider the different deviation scales of voltage and frequency. The developed graphical tool-based design methodology may not provide an optimal choice for the LS synthesis; however, it presents a useful trade-off between simplicity and preciseness. Further research is needed to improve the proposed LS strategy.

8.4 WAVE PROPAGATION-BASED EMERGENCY CONTROL

Power system angle instability following loss of synchronism of the generators can be considered as a fast instability phenomena [6,26]. Detection of this phenomenon and performing adequate emergency actions are important issues to maintain power system stability. In this section, the concept of electromechanical wave propagation is used to perform an effective emergency control scheme in a large-scale power system.

Here, a new power system emergency control framework based on the descriptive study of electrical measurements and electromechanical wave propagation (described in Section 4.6) is introduced. Since fast and accurate detection of instability is essential in initiating certain emergency control measures, the proposed methodology could be also useful in detecting the contingency condition and performing the well-known islanding and LS techniques. The section is supplemented by some illustrative nonlinear simulations on the large-scale test systems.

8.4.1 Proposed Control Scheme

The overall framework of the proposed control methodology is demonstrated in Fig. 8.14. This figure summarizes the process of using islanding formation, performing a continuum model for system stability assessment, and predicting suitable emergency actions following a contingency.

Angle instability is a fast instability phenomenon. Therefore, predicting its situation and performing suitable actions are very important. As mentioned in Section 4.6, having a continuum model of a power system can be helpful for predicting the trajectory of the disturbances by using disturbance conditions as initial states of the continuum model. Here, the power system continuum model is used to provide a powerful descriptive tool for stability analysis in emergency conditions.

The slow coherency theory can be used to identify the system islands. Determining the islands leads to identification of the weak links or critical cutset. Having knowledge of the weak links/islands helps one to determine the most suitable connections/locations for performing a more careful islanding plan, when the system needs to be separated. Here, a slow coherency is suggested to find the weak links/islands. The weak links are used to check whether islanding is needed or not. The weak links can be considered as those links that should be tripped when the islanding is recommended.

It is noteworthy that the overall stability can be evaluated by monitoring just a few links, that is, the weak links and the links near the contingency location. Furthermore, observing the trajectory of bus voltage angles helps the system operators to choose a suitable islanding plan. Following an islanding action, the power system is divided into some islands with excess load/generation. Therefore, other emergency control actions [13,27] such as the LS and generation tripping should be performed, as shown in Fig. 8.14.

Indeed, following a certain contingency, the most important goal of the proposed algorithm is to determine if islanding is needed or not. If the contingency is not dangerous

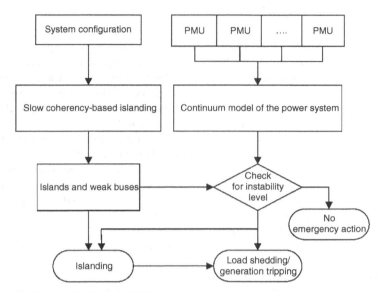

Figure 8.14. Overall framework of the proposed methodology.

and the angle across a link does not exceed a certain value (e.g., 30° for noncompensated lines), islanding is not needed.

However, for the higher value of transmission angles, the other emergency action will not be able to restore system stability. For more clarification, assume the angle across a link is increasing and exceeds 90°. After that, a decrease of the active power, for example, using LS, could not restore the system because the angle may track the power at the low side of the power–angle (P–δ) curve. Therefore, δ increases and thus instability occurs. In these circumstances, the islanding is strongly needed. The critical angle values (thresholds) should be determined based on the specified level of security.

Operators and engineers can validate the power system stability at the monitoring/control (SCADA) center by observing the wave propagation at the human–machine interface (HMI). Three modes may be defined for this tool: real-time, prediction, and test modes.

In real-time mode, the real-time data gathered from the network by the PMUs are shown as a surface. The operator can see the real-time states of the whole network. In prediction mode, the online data are used as initial values of a continuum model, and then the system states at a certain time value will be predicted. In the test mode, the operator or engineer can validate a certain contingency based on the real states of the system. In this mode the real-time data are used as initial values of a continuum model and a certain test is used as a deviation from initial state; then the post contingency condition will be shown in HMI for a certain time interval.

8.4.2 Simulation Results

8.4.2.1 Ring Systems and Islanding. To illustrate the concept of a coherent group of generators following a contingency, a 200-bus ring system is simulated. For simplicity, it is assumed that there is only one generator or one load at each bus. All generators are similar with equal amount of power and all loads are also equal. The number of generators (N_G) is equal to the number of loads (N_L), so $N_G = N_L = 100$. The system data are determined the same as considered in Ref. [27].

The system is examined under two different configurations. In the first configuration (config-1), all generators and loads are distributed throughout the power system by a uniform random function. In the second configuration (config-2), the generators and loads are distributed in a three-area ring system, as shown in Fig. 8.15. In this case, all line impedances are assumed to be fixed at 0.1 pu, except three lines where their impedances are $Z^{75-76} = 0.2$, $Z^{130-131} = 0.15$, $Z^{200-1} = 0.3$. For both configurations, a large disturbance, that is, tripping line 200–1, is applied. The angle deviations are illustrated in Fig. 8.16. The ring system is opened due to occurred disturbance.

As shown in Ref. [27], for the unstable case (config-2), the angle across link 75–76, that is, $\delta^{75-76} = \delta^{75} - \delta^{76}$, is continually growing and finally this situation leads toward separation and instability. However, for the stable case (config-1), although the angle across the link deviates, the system remains in a limited boundary and moves to a constant value. The kinetic energy, potential energy, and total energy across link 75–76 are also given in Ref. [27]. As already mentioned, the system will be stable if it can be able to convert all amount of its kinetic energy achieved during a contingency into potential energy. This simulation also shows that following the

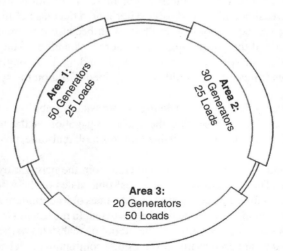

Figure 8.15. 200-Bus ring system (config-2).

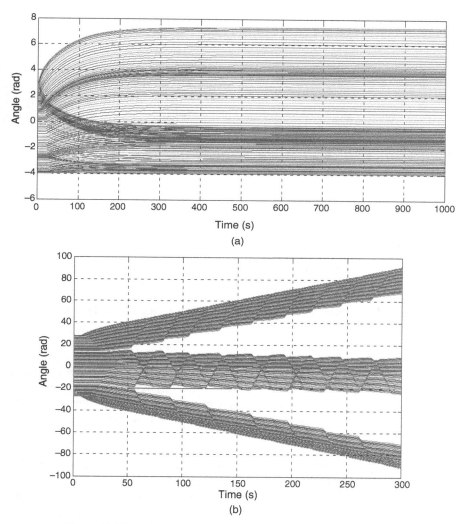

Figure 8.16. Wave propagation for: (a) config-1 and (b) config-2.

mentioned fault, config-2 goes to an unstable condition, while config-1 maintains its stability.

As another example, assume a Gaussian distribution that affects the angles of a system depicted in Fig. 8.15 (config-2). Postcontingency wave propagation is illustrated in Fig. 8.17. In this case, the center of disturbance is located at bus 60; however, it can be seen that the system is separated at line 75–76, which is a weak link. Actually, when a disturbance reaches a weak link through its propagation trajectory, this may lead to an unstable operating point.

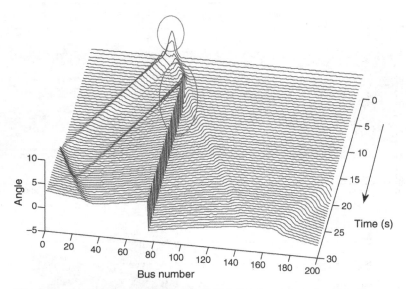

Figure 8.17. Wave propagation (3D plot) following a Gaussian disturbance.

Figure 8.17 clearly shows the behavior of wave propagation when it reaches a weak link. For plotting this figure, the angle variations versus bus number for each time slot is calculated. As can be seen, angle wave is reflected when it reaches a transmission line without enough stability margin, and the system is separated exactly at the weak point.

8.4.2.2 *Application to 24-Bus Test System.* Here, the well-known 24-bus reliability test system (Fig. 8.8) is used to investigate the effectiveness of the proposed strategy. To make a serious fault, the connections between area 1 and area 2 are loosened. Now, the angle instability on link 24–3 can be considered as a useful example to examine the proposed methodology. Assume lines 16–19 and 16–14 are disconnected at $t = 2$ s. Following this large disturbance, line 3–24 will encounter a high overloading problem. This overloading is larger than its angle stability limit. Therefore, as shown in Ref. [27], the angles of generators G1 and G15 located at two sides of the line are separated, and the angle instability phenomena immediately occurs. To save the system, one solution is the use of islanding control.

Figure 8.18 illustrates how to use the proposed control scheme; an islanding plan improves the voltage behavior that has been suddenly depressed following the mentioned event. Therefore, the remaining tie-lines may be tripped when appropriate algorithms are not used to stabilize the resulting two islands. To save an island with excess load, a UFLS algorithm, like those suggested by the FRCC or addressed in

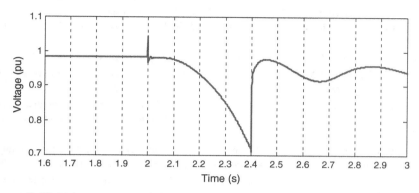

Figure 8.18. Voltage response following disturbance (at 2 s) and islanding plan (at 2.4 s).

Refs [6,10,13,19] should be used. However, for the islands with excess generation, some loads must be switched on.

8.5 SUMMARY

The UFLS and UVLS schemes are usually in use in real-world power systems without coordination. In the present chapter, the necessity of using both voltage and frequency data, specifically in the presence of high wind power penetration, to develop an effective LS scheme is emphasized.

In the first part, it is shown that the voltage and frequency responses may behave in opposite directions following many contingencies. The introduced graphic tool (in Chapter 4) to study the system dynamic frequency and voltage behavior in emergency conditions is used to propose a new LS scheme using both voltage and frequency information. The present study and given statements are supported by performing nonlinear simulations on the 9-bus and 24-bus IEEE test systems.

In the second part, based on the given descriptive study of electrical measurements and electromechanical wave propagation in large electric power systems, an emergency control scheme is introduced to detect the possible plans.

REFERENCES

1. IEEE Standard C37.117, IEEE Guide for the Application of Protective Relays Used for Abnormal Frequency Load Shedding and Restoration, 2007, pp. c1–c43.
2. UVLS Task Force Technical Studies Subcommittee, Undervoltage Load Shedding Guideline, WSCC, July 1999.
3. C. Mozina, Undervoltage load shedding: part 1, Electric Energy T&D Magazine, May–June 2006.

4. S. H. Mark, A. H. Keith, A. J. Robert, and Y. T. Lee, Slope-permissive under-voltage load shed relay for delayed voltage recovery mitigation, *IEEE Trans. Power Syst.*, **23**(3), 1211–1216, 2008.

5. H. Bevrani, *Robust Power System Frequency Control*, Springer, New York, 2009.

6. P. M. Anderson and M. Mirheidar, An adaptive method for setting underfrequency load shedding relays, *IEEE Trans. Power Syst.*, **7**(2), 647–655, 1992.

7. H. Bevrani, G. Ledwich, and J. J. Ford, On the use of df/dt in power system emergency control. In: *Proceedings of the IEEE Conference & Exposition, Power Systems*, Seattle, WA, March 15–18, 2009.

8. H. You, V. Vittal, and Z. Yang, Self-healing in power systems: an approach using islanding and rate of frequency decline-based load shedding, *IEEE Trans. Power Syst.*, **18**(1), 174–181, 2003.

9. J. J. Ford, H. Bevrani, and G. Ledwich, Adaptive load shedding and regional protection, *Int. J. Electr. Power*, **31**, 611–618, 2009.

10. H. Bevrani, G. Ledwich, Z. Y. Dong, and J. J. Ford, Regional frequency response analysis under normal and emergency conditions, *Electr. Power Syst. Res.*, **79**, 837–845, 2009.

11. H. Bevrani and T. Hiyama, On load-frequency regulation with time delays: design and real time implementation, *IEEE Trans. Energy Convers.*, **24**(1), 292–300, 2009.

12. V. T. Vladimir, Adaptive under-frequency load shedding based on the magnitude of the disturbance estimation, *IEEE Trans. Power Syst.*, **21**(3), 1260–1266, 2006.

13. H. Bevrani and A. G. Tikdari, An ANN-based power system emergency control scheme in the presence of high wind power penetration. In: L. F. Wang, et al., editors, *Wind Power Systems: Applications of Computational Intelligence*, Springer Book Series on Green Energy and Technology, Springer, Heidelberg, 2010.

14. H. Bevrani and A. G. Tikdari, On the necessity of considering both voltage and frequency in effective load shedding schemes. In: *Proceedings of the IEEJ Technical Meeting*, PSE-10-02, Fukui, Japan, January 21, 2010, pp. 7–11.

15. D. Gautam, V. Vittal, and T. Harbour, Impact of increased penetration of DFIG-based wind turbine generators on transient and small signal stability of power systems, *IEEE Trans. Power Syst.*, **24**(3), 1426–1434, 2009.

16. L. Gillian, M. Alan, and O. M. Mark, Frequency control and wind turbine technologies, *IEEE Trans. Power Syst.*, **20**(4), 1905–1913, 2005.

17. M. V. A. Nunes, J. A. Peças Lopes, H. H. Zurn, U. H. Bezerra, and R. G. Almeida, Influence of the variable-speed wind generators in transient stability margin of the conventional generators integrated in electrical grids, *IEEE Trans. Energy Convers.*, **19**(4), 692–701, 2004.

18. C. Yongning, L. Yanhua, W. Weisheng, and D. Huizhu, Voltage stability analysis of wind farm integration into transmission network. In: *IEEE International Conference on Power System Technology*, Chongqing, October 22–26, 2006, pp. 1–7.

19. A. G. Tikdari, Load shedding in the presence of renewable energy sources, Dissertation, University of Kurdistan, Sanandaj, Iran, 2009.

20. P. M. Anderson and A. A. Fouad, *Power System Control and Stability*, IEEE Press, New York, 2003.

21. PRC-006-FRCC-01, FRCC Automatic Underfrequency Load Shedding Program, 2009. Available at https://www.frcc.com/.

22. IEEE RTS Task Force of APM Subcommittee, IEEE reliability test system, *IEEE Trans. Power Apparatus Syst.*, **98**(6), 2047–2056, 1979.

23. IEEE RTS Task Force of APM Subcommittee, The IEEE reliability test system-1996, *IEEE Trans. Power Syst.*, **14**(3), 1010–1020, 1999.

24. D. L. H. Aik, A general-order system frequency response model incorporating load shedding: analytic modeling and application, *IEEE Trans. Power Syst.*, **21**(2), 709–717, 2006.

25. Y. Halevi and D. Kottick, Optimization of load shedding system, *IEEE Trans. Energy Convers.*, **8**(2), 207–213, 1993.

26. P. Kundur, *Power System Stability and Control*, McGraw-Hill, New York, 1994.

27. A. G. Tikdari, H. Bevrani, and G. Ledwich, A descriptive approach for power system stability and security assessment. In: P. Vasant, N. Barsoum, and J. Webb, editors, *Innovation in Power, Control and Optimization: Emerging Energy Technologies*, IGI Global, Hershey, PA, 2011.

MICROGRID CONTROL: CONCEPTS AND CLASSIFICATION

Currently, economical harvesting of electrical energy on a large scale considering the environmental issues is undoubtedly one of the main challenges. As a solution micro/smart grids promise to facilitate the wide penetration of renewable energy sources (RESs) and energy storage devices into the power systems, reduce system losses and greenhouse gas emissions, and increase the reliability of electricity supply to customers. Due to their potential benefits of providing secure, reliable, efficient, sustainable, and environmentally friendly electricity from RESs, the interest on micro/smart grids is growing.

Microgrids (MGs), as basic elements of future smart grids, have an important role in increasing the grid efficiency, reliability, and satisfying the environmental issues. Although the concept of microgrids is already established, the control strategies and energy management systems for microgrids, which cover power interchange, system stability, frequency and voltage regulation, active and reactive power control, islanding detection, grid synchronization, and system recovery are still under development. In this chapter, a comprehensive review on various microgrid control loops and relevant standards are given with a discussion on the challenges of microgrid controls. A summary of this chapter has already been presented in Ref. [1].

Power System Monitoring and Control, First Edition. Hassan Bevrani, Masayuki Watanabe, and Yasunori Mitani.
© 2014 John Wiley & Sons, Inc. Published 2014 by John Wiley & Sons, Inc.

9.1 MICROGRIDS

A microgrid (MG) is an interconnection of domestic distributed loads and low-voltage (LV) distributed energy sources, such as microturbines, wind turbines, photovoltaics (PV), and storage devices. The microgrids are placed in the low voltage (LV) and medium voltage (MV) distribution networks. This has important consequences. With numerous microsources connected at the distribution level, there are new challenges, such as system stability, power quality, and network operation that must be resolved by applying the advanced control techniques at LV/MV levels rather than high-voltage levels, which is common in conventional power system control. In other words, distribution networks (demand side) must pass from a passive role to an active one.

A simplified MG architecture is shown in Fig. 9.1. This MG consists of a group of radial feeders as a part of a distribution system. The domestic load can be divided to sensitive/critical and nonsensitive/noncritical loads via separate feeders. The sensitive loads must be always supplied by one or more microsources, while the nonsensitive loads may be shut down in case of contingency or a serious disturbance.

Each unit's feeder has a circuit breaker and a power flow controller commanded by the central controller or energy manager. The circuit breaker is used to disconnect the corresponding feeder (and associated unit) to avoid the impacts of severe disturbances through the MG. The MG is connected to the distribution system by a point of common coupling (PCC) via a static switch (SS; Fig. 9.1). The static switch is capable of islanding the MG for maintenance purposes or when a fault or contingency occurs. All such events are well described in the IEEE 1547 standard [2].

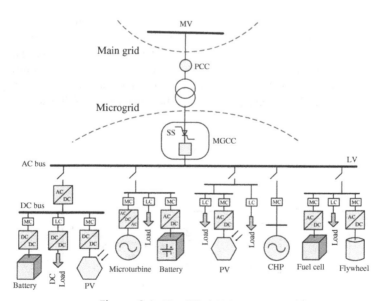

Figure 9.1. Simplified MG structure.

For the feeders with sensitive loads, local power supply, such as diesel generators or energy capacitor systems (ECSs) with enough energy-saving capacity are needed to avoid interruptions of the electrical supply. The MG central controller (MGCC) [3] facilitates a high-level management of the MG operation by means of technical and economical functions. The microsource controllers (MCs) control the microsources and the energy storage systems. Finally, the controllable loads are controlled by the load controllers (LC).

The microsources and storage devices use power electronic circuits to connect to the MG. Usually, these interfaces depending on the type of unit and connected feeder are AC/AC, DC/AC, and AC/DC power electronic converters/inverters. As the MG elements are mainly power-electronically interfaced, the MG control depends on the inverter control.

There are a variety of modulation techniques that can be used in power electronic inverters/converters including pulse width modulation (PWM), hysteresis modulation, and pulse density modulation (PDM). Hysteresis modulation is perhaps the simplest, but due to some shortcomings in providing high-quality output current and good transient response, it is not preferred for the MG inverters. The PWM is the most common modulation technique in the MG's inverters/converters. The PDM technique is another possible modulation technique, which is used in high-frequency converters applied for induction heating applications.

Generally, the inverters have two separate operation modes, acting as a current source or as a voltage source. The general model for an inverter-based microsource is shown in Fig. 9.2. A microsource contains three basic elements: power source or prime mover, DC interface, and inverter. The microsource couples to the MG through a power line. The output voltage and frequency, as well as real and reactive powers of the microsource can be controlled using local feedback applied to the inverter.

In comparison to the conventional generators, the microsources (DGs) such as natural gas and diesel generating units are very fast and can typically pick up load within 10–12 s from startup and can serve full load just a few seconds thereafter. The microsource can control the phase and magnitude of its output voltage V and from the line reactance X; it can determine the transferring real power P and reactive power Q flows from itself to the grid. The P and Q values can be calculated as follows:

$$P = \frac{3}{2}\frac{VE}{X}\sin\delta \qquad (9.1)$$

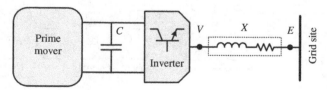

Figure 9.2. A model for a microsource connected to a MG.

$$Q = \frac{3}{2}\frac{V}{X}(V - E \cos \delta) \qquad (9.2)$$

where

$$\delta = \delta_V - \delta_E \qquad (9.3)$$

The E is the voltage at the grid side of the connecting line; the δ_V and δ_E are the angles of V and E, respectively. For small δ, the P and Q mainly depend on the δ and V, respectively:

$$P \approx \frac{3}{2}\frac{VE}{X}\delta \qquad (9.4)$$

$$Q \approx \frac{3}{2}\frac{V}{X}(V - E) \qquad (9.5)$$

These relationships allow us to establish feedback loops in order to control output power and MG voltage in islanding.

These relationships show that if the reactive power in the MG generated by the microsources increases, the local voltage must decrease, and vice versa. Also, there is a similar behavior for frequency versus real power. These relationships, which are formulated in (9.6) and (9.7), allow us to establish feedback loops in order to control MG's real/reactive power and frequency/voltage.

$$\omega - \omega_0 = -R_P(P - P_0) \qquad (9.6)$$

$$V - V_0 = -R_Q(Q - Q_0) \qquad (9.7)$$

The RP and RQ are known as the frequency and voltage droop coefficients.

The ω_0, P_0, V_0, and Q_0 are the nominal values (references) of frequency, active power, voltage, and reactive power, respectively. A graphical representation for (9.6) and (9.7) is shown in Fig. 9.3.

The interconnected DG units with different droop characteristics can jointly track the load change to restore the nominal system frequency and voltage. This is illustrated in Fig. 9.3, representing two units with different droop characteristics connected to a common load. The DGs are operating at a unique nominal frequency/voltage with different output active/reactive powers. The change in the network load causes the microsources to decrease their speed/voltage, and hence, the units increase the output powers until they reach a new common operating frequency/voltage. As expressed in (9.8), the amount of produced power by each DG to compensate the network load change depends on the unit's droop characteristics [4].

$$\Delta P_{gi} = \frac{\Delta \omega}{R_{Pi}}, \Delta Q_{gi} = \frac{\Delta V}{R_{Qi}} \qquad (9.8)$$

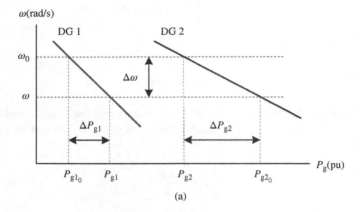

(a)

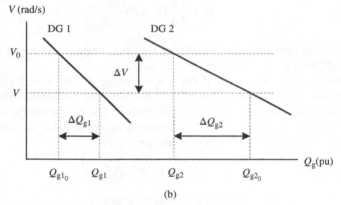

(b)

Figure 9.3. Droop control characteristics: (a) $\omega - P$ droop and (b) $V - Q$ droop.

Hence,

$$\frac{\Delta P_{g1}}{\Delta P_{g2}} = \frac{R_{P2}}{R_{P1}} \tag{9.9}$$

and

$$\frac{\Delta Q_{g1}}{\Delta Q_{g2}} = \frac{R_{Q2}}{R_{Q1}} \tag{9.10}$$

It is noteworthy that the described droop control characteristics in (9.6), (9.7), and Fig. 9.3 have been obtained for electrical grids with inductive impedance ($X \gg R$) and a great amount of inertia, which is the case in conventional power systems with high voltage lines. In a conventional power system, immediately following a power imbalance due to a disturbance, the power is going to be balanced by natural response generators using rotating inertia in the system via the primary frequency control loop [5]. In the MG on the other hand, there is no significant inertia and if an unbalance occurs between the generated power and the absorbed power, the voltages of the power sources change. Therefore, in this case, voltage is triggered by the power changes.

In fact, for medium and low voltage lines that the MGs are working with, the impedance is not dominantly inductive ($X \cong R$). For resistive lines, reactive power Q mainly depends on δ and real power P depends on voltage V [6]. This fact suggests different droop control characteristics, called opposite droops. Recently, several research works have been done to introduce new and specific droop characteristics for the MG control design purposes.

However, each microgenerator has a reference reactive power to obtain a voltage profile that matches the desirable real power. In low-voltage grids, Q is a function of δ, which is adjusted with the V versus P droop. It means that there is a possibility to vary the voltage of generators exchanging the reactive power [7,8]. Therefore, the conventional droops are still operable in low-voltage grids and MGs.

In a grid-connected operation, MG loads receive power from both grid and local microsources, depending on the customer's situation. In emergency conditions, for example, following a problem for the main grid (such as voltage drops, faults, blackouts), the MG can be separated from the grid via a static switch in about a cycle, as smoothly as possible. The MG can be also islanded intentionally for specific reasons even though there is no disturbance or serious fault in the main grid side. In these cases, the MG operation continues in islanding operation mode.

The balance between generation and demand of power is one of the most important requirements in MG management in both grid-connected and islanded operation modes. In the grid-connected mode, the MG exchanges power with an interconnected grid to meet the balance, while, in the islanded mode, the MG should meet the balance for the local supply and demand using the decrease in generation or load shedding.

During the grid-connected mode, the generating units operate in current-control mode, in which they should regulate the exchange of active and reactive powers between the MG and the main grid. During islanded operation, the DGs operate in voltage-control mode to regulate the MG voltage and to share the local loads. In islanding, if there are local load changes, local microsources will either increase or decrease their production to keep constant the energy balance, as much as possible. In an islanded operation, an MG works autonomously; therefore, it must have enough local generation to supply demand, at least to meet the sensitive loads. This is not the case in grid-connected operation, because in this situation, the main grid compensates the increases or decreases of the load.

The islanding operation could happen under two scenarios: planned (intentional) and unplanned (unintentional) islanded operations. The planned islanded operation can be done for maintenance purposes, economical criterion, or in case of long-term voltage dips or general faults following an event in the main grid. The unplanned islanded operation may happen following a contingency such as severe disturbance (or blackout) in the main grid.

Immediately after islanding, the voltage, phase angle, and frequency at each microsource in the MG change. For example, the local frequency will decrease if the

MG imports power from the main grid in the grid-connected operation, but will increase if the MG exports power to the main grid in the grid-connected operation.

9.2 MICROGRID CONTROL [1]

The main profits associated with the MG concept can be considered as efficiency improvement in energy transmission, considerable reduction in environmental pollution (e.g., emissions of CO_2 and SO_2), and security/reliability enhancement, considering the inherent redundancy of DGs. However, the high penetration of DGs certainly increases the complexity of control, protection, and communication of distribution systems, which are mainly designed to operate radially without any generation at the low-voltage distribution lines or customer side. An important issue is how to integrate the numerous MGs into the existing distribution networks by properly coordinating their generator/ storage units operation and by limiting their potentially negative side effects on network operation and control.

Control is one of the key enabling technologies for the deployment of MG systems. The MG has a hierarchical control structure with different layers. The MGs require effective use of advanced control techniques at all levels. The secure operation of MGs in connected and islanding operation modes, as well as successful disconnection or recon-nection processes depend upon MG controls. The controllers must guarantee that the processes occur seamlessly and the system is working in the specified operating points.

Due to the high diversity in generation and loads, the MGs exhibit high non-linearities, changing dynamics, and uncertainties that may require advanced robust/ intelligent control strategies to solve. The use of more efficient control strategies would increase the performance of these systems. Since some RESes such as wind turbines and PVs are working under turbulent and unpredictable environmental conditions, the MGs have to adapt to these variations, and in this way the efficiency and reliability of MGs strongly depend on the applied control strategies.

As already mentioned, the MGs should be able to operate autonomously but also interact with the main grid. In connected operation mode, the MGs are integrated to a constantly varying electrical grid with changing tie-line flow, voltages, and frequency. To cope with these variations and to respond to grid disturbances—performing active power/frequency regulation and reactive power/voltage regulation—the MGs need to use proper control loops. Furthermore, suitable islanding detection feedback/algorithms are needed for ensuring a smooth transition from grid-connected to islanded mode to avoid cascaded failures.

In islanded mode, the MG operates according to the existing standards (e.g., IEEE 1547) and the existing controls must properly work to supply the required active and reactive power as well as to provide voltage and frequency stability. A controlled switch reconnects the MG to the grid when the grid voltage is within acceptable limits and the phasing is correct. In this stage, active synchronization is required to match the frequency, voltage, and phase angle of the MG.

A general scheme for operating controls in an MG is shown in Fig. 9.4. Each MG is locally controlled by the MCs. The LCs are installed at the controllable loads to provide

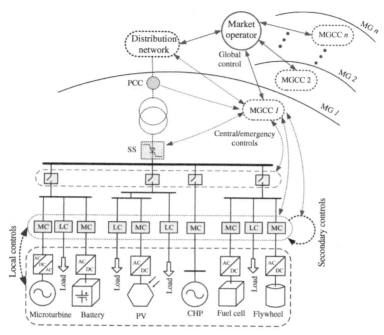

Figure 9.4. A general scheme for MG controls.

load control capabilities. For each MG, there is a central controller (MGCC) that interfaces between the distribution management system (DMS) or distribution network operator (DNO) and the MG. The DMS/DNO has the responsibility to manage the operation of medium- and low-voltage areas in which more than one MG may exist. Later, these controllers are explained in detail.

Similar to the conventional power systems [9], the MGs can operate using various control loops, which can be mainly classified into four control groups: local, secondary, central/emergency, and global controls. The *local control* deals with initial primary control such as current and voltage control loops in the microsources. The *secondary control* ensures that the frequency and average voltage deviation of the MG is regulated toward zero after every change in load or supply. It is also responsible for inside ancillary services. The *central/emergency control* covers all possible emergency control schemes and special protection plans to maintain the system stability and availability in the face of contingencies. The emergency controls identify proper preventive and corrective measures that mitigate the effects of critical contingencies. The *global control* allows MG operation at an economic optimum and organizes the relation between an MG and distribution network as well as other connected MGs.

In contrast to the local control, operating without communication, secondary, global, and emergency controls may need communication channels. On the other hand, the local controls are known as *decentralized* controllers, the global, and to some extent, secondary and emergency controllers are operating as *centralized* controllers.

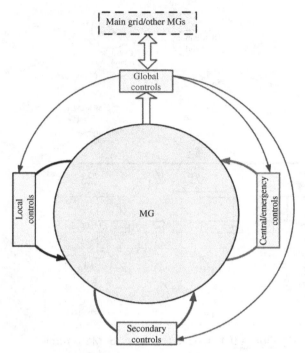

Figure 9.5. MG controls.

Figure 9.5 shows a conceptual framework for the described operating control loops in an MG. In summary, existing MG's control loops in the four mentioned groups have the following responsibilities:

1. Working of all microsources at the predefined operating points,
2. Interchanging active and reactive powers according to the scheduled plan,
3. Meeting the operating limits by all important electrical indices such as voltage and frequency among the MG,
4. Seamlessly islanding and resynchronizing processes using proper techniques,
5. Market participation optimizing,
6. Reducing the circulating currents among parallel connected microsources/inverters,
7. Guarantee secure power supply for sensitive loads,
8. Capability of operation through black start in case of general failure,
9. Providing emergency control and protective schemes such as load-shedding,
10. Possibility of remote operation of circuit breakers, and
11. Proper using of energy storage devices.

9.3 LOCAL CONTROLS

Local or internal controls appear in different forms depending on the type of micro-sources that can be addressed based on their technologies such as induction generators, synchronous generators, and power electronic inverters/converters. Some microsources such as fuel cells and PV cells generate DC power, which for operation in an AC MG, must be connected to the network through the DC/AC converters.

Older wind turbines and small hydro units use fixed speed induction generatorss (FSIGs) that are connected directly to the grid. Modern variable speed wind turbines use doubly fed induction generators (DFIGs) with their stators connected directly to the grid and their rotors connected via AC/DC–DC/AC converters. Some other power sources, such as combined heat and power (CHP) units and microturbines use synchronous generators. Synchronous generators operate at their synchronous speed if they are directly connected, similar to the control of large conventional generating units.

In the FSIG wind turbines, the active power is merely determined by the mechanical power input, but reactive power and power factor can only be controlled with shunt compensators [10]. However, in the DFIG wind turbines, the rotor side converter controls the reactive power flow either for voltage or power factor control, and sets the rotor voltage and frequency for maximum power point tracking (MPPT). The grid side converter controls the power flow in order to maintain the DC-link capacitor voltage [11].

In comparison with synchronous and induction generating units, the power electronic inverters/converters provide more flexible operation. The source-side inverter is usually a voltage source inverter (VSI) and is controlled to provide MPPT in wind turbine applications. The grid-side inverter in the role of a line commutated inverter or a VSI controls the DC-link voltage to provide the MPPT for PV or wind turbines with synchronous generators and diode rectifiers [12], and also it can control the active and reactive power output.

The local controls deal with the inner control of the DG units that usually do not need the communication links resulting in simple circuitry and low cost. Local controls are the basic category of the MG controls. The main usage of local controllers is to control microsources (Fig. 9.2) to operate in normal operation. This type of controllers is aimed at controlling the operating points of the microsources and their power-electronic interfaces.

These controls are going to be more vital for an MG due to integration of a large number of microsources in order to overcome fluctuation caused by the high penetration of microsources. Some loads can be also locally controllable using the LCs. The LCs are usually used for demand side management.

For example, in solar plants the local controls are related to sun tracking and control of the thermal variables. Although control of the sun-tracking mechanisms is typically done in an open-loop mode, control of the thermal variables is mainly done in the closed loop mode. In microturbines and inverter-based energy sources such as wind turbines and uninterruptible power supply (UPS)-based energy storage systems, it is the droop control that ensures that the active and reactive powers are properly shared between the inverters. The local control loops are also responsible for regulating the unit output voltage and limit the output current.

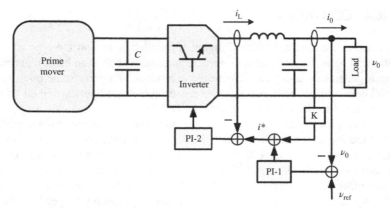

Figure 9.6. Local controls for a stand-alone inverter-based DG.

The main function of a DG in stand-alone and islanded mode is to assure system stability and desirable performance by providing correct voltage and frequency in order to supply the local load. Figure 9.6 depicts a block diagram of local control loops for stand-alone inverter-based microsources. The outer loop regulates the output capacitor voltage v_0. After the addition with the measured output current, it sets the reference inductor current i^* for the inner control loop. Blocks PI-1 and PI-2 are the voltage and the current based proportional-integral (PI) regulators, respectively. The voltage and frequency of the filter output voltage reference signal v_{ref} are kept constant, but their values could vary in case of working in the grid-connected operation mode; in this state, additional control, that is, droop control, should be used.

Besides the voltage and frequency controls, microsources must control active and reactive powers. The droop-based active and reactive power controls are the most common methods to control these powers. As described in Section 9.1, these droop controls are similar to the existing versions of droop-based controls in the conventional power systems. The droop-based control depicts the relation voltage and reactive power $(Q–V)$, as well as frequency and active power $(P–\omega)$ indices. Figure 9.7a shows a simple realization for droop-based control loops (9.6) and (9.7) from output current and voltage measurements. As shown in Fig. 9.7b, the results can be used to provide the inverter voltage reference.

As the generated reactive power by the microsource increases (becomes more capacitive), the operating voltage increases, too. Therefore, the local voltage set-point should be reduced to keep the voltage at or near its nominal set-point. The same behavior exists for frequency and active power. A typical droop-based multiloop local control structure is shown in Fig. 9.8.

In the case of parallel inverters, these control loops, also called $P–\omega$ and $Q–V$ droops, use feedback from the voltage and frequency of each microsource/inverter for sensing the output average active and reactive powers to emulate virtual inertias. Therefore, in power electronic-based MGs, the droop control can be done by adding

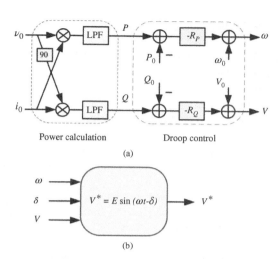

Figure 9.7. Realization of droop characteristics.

virtual inertias and controlling the output impedances; this can be useful in controlling active and reactive power injected to the grid. However, in the last case, the droop control faces several challenges that should be solved using advanced control methodologies. A slow transient response, line impedance dependency, and poor active/reactive power regulation are some of these challenges.

Synthesis of the local MG controllers is a crucial issue. The local controllers' design should be based on a detailed dynamic model of the MG, including the resistive, reactive, and capacitive local load and the distribution system. This model should be adapted to the practical operating conditions of the MG in order to guarantee that the controllers respond properly to the system's inherent dynamics and transients [13]. Some local control design examples are given in Refs [14–17].

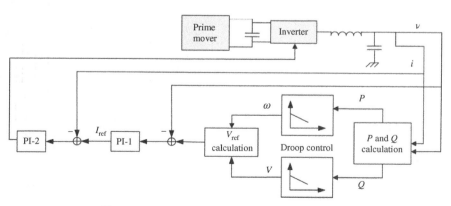

Figure 9.8. Droop-based multiloop local control.

9.4 SECONDARY CONTROLS

Secondary controls as the second layer control loops complement the task of inner control loops to improve the power quality inside the MG and to enhance the system performance by removing the steady-state errors. They are closely working with local and global control groups.

During the grid-connected operation, all the microsources and inverters in the MG use the grid electrical signal as reference for voltage and frequency. Since in this mode, the active (P) and reactive (Q) powers are controlled by the main grid, this control mechanism is known as P–Q control. Figure 9.9 shows a typical control structure for the P–Q control.

However, in islanding, the DGs lose the reference signal provided by the main grid. In this case they may coordinate to manage the simultaneous operation using one of following secondary control methods:

1. *Single Master Operation:* a master microsource/inverter fixes voltage and frequency for the other units in the MG. The connected microsources are operating according to the reference given by the master (Fig. 9.10).
2. *Multimaster Operation:* in this case, several microsources/inverters are controlled by means of a central controller such as MGCC, which chooses and transmits the set points to all the generating units in the MG [18].

Secondary controls also cover some of the controls needed to improve the parallel operation performance for DGs (or inverters). Sometimes, the commands provided by these controls are distributed through a low-bandwidth communication channels to the

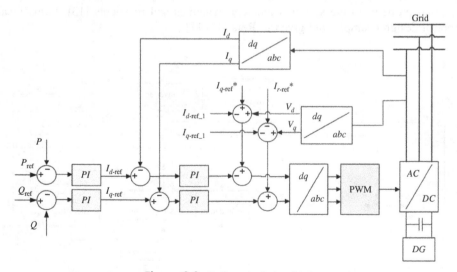

Figure 9.9. *P–Q* control structure.

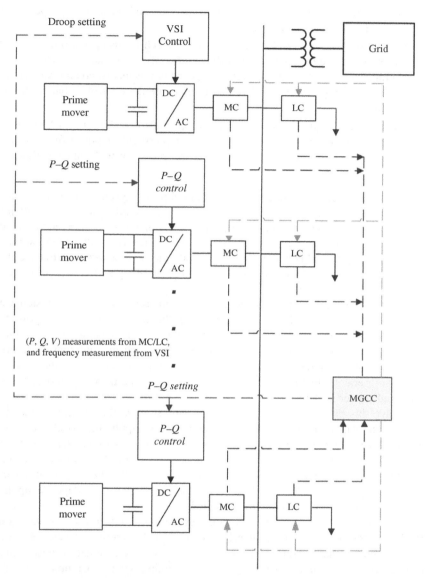

Figure 9.10. Single master operation control structure.

parallel DGs/inverters. There are many control techniques in the literature to make a successful parallel operation of DGs/inverters; they can be categorized into three main approaches [19]:

1. Master/slave control techniques, which use a voltage-controlled inverter as a master unit and current-controlled inverters as the slave units [20]. The master

unit maintains the output voltage sinusoidal and generates proper current commands for the slave units.

2. Current/power sharing control techniques, which by using them the total load current is measured and divided by the number of units in the system to obtain the average current. The actual current from each unit is measured and the difference from the average value is calculated to generate the control signal for the load sharing.

3. Generalized frequency and voltage droop control techniques, which use the normal conventional frequency/voltage droop control, opposite frequency/voltage droop control, or a combination of droop control with other methods.

Similar to the secondary control in conventional power systems, secondary controls in MGs are responsible of providing ancillary services. According to the IEEE Standard 1547 [2], the ancillary services in distributed power generation systems are defined as load regulation, energy losses, spinning and nonspinning reserve, voltage regulation, and reactive power supply. This standard recommends that low-power systems should be disconnected when the grid voltage is lower than 0.85 pu or higher than 1.1 pu as an anti-islanding requirement [2,21].

In the MGs, because of the variable nature of some renewable energy systems such as PV or wind turbine, and difficulty in predicting the amount of produced power, the peaks of power demand may not necessarily coincide with the generation peaks. On the other hand, a network of small-sized microsources, which are dominated by power electronic-interfaced sources, do not have enough inertia to respond to the initial and surge power or energy mismatch by using their machines' inertia as commonly found in conventional power systems.

To solve this problem, storage energy systems such as flow batteries, fuel cells, flywheels, and superconductor inductors are used to supply the local loads in an uninterruptible manner. These storage devices could be also useful to support regulation tasks and ancillary services in coordination with the MG's DGs. Coordination of storage devices and DGs for providing ancillary services to improve the system performance can be considered as a secondary control. The capacity of the ECS depends upon the characteristics of the regulation being provided.

An experimental control design example for using of ECS in a multiagent system (MAS)-based coordination with a diesel generator for the load-frequency control as a secondary control issue is described in Ref. [22]. The MG is considered as an isolated grid with dispersed microsources such as photovoltaic units, wind generation units, diesel generation units, and an ECS for the energy storage. The addressed scheme has been proposed through the coordination of controllable power microsources such as diesel units and the ECS with small capacity. All the required information for the proposed frequency control is transferred between the diesel units and the ECS through computer networks. The applied control structure is shown in Fig. 9.11. In this figure, W_{ECS} and P_{ECS} are the current stored energy and the produced power by the ECS unit, respectively.

Here, ΔP_{ECS} and ΔP_{DG} represent the control action signals for output setting of ECS and diesel unit, respectively. Applying the control signal ΔP_{ECS} provides an appropriate

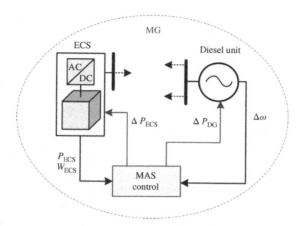

Figure 9.11. MAS-based coordinated ECS-diesel generator frequency control in MG.

charging/discharging operation on the ECS for the frequency regulation purpose. Because of the specific feature of the ECS dynamics, the fast charging/discharging operation is possible to achieve in an ECS unit. Therefore, the variations of power generation from the wind turbine and PV units, and in addition, the variation of demand power on the variable loads, can be efficiently absorbed through the charging/discharging operation of the ECS unit. An additional regulation power (from the diesel units) is required to keep the stored energy level of the ECS in a proper range.

Figure 9.12 illustrates the dynamic configurations of the coordinated control loops for the diesel unit and ECS located in the MAS control unit. In this study, the communication time delay is also considered. ΔW_{ref} and ΔW_{ECS} are the target and measured available energies in the ECS. The ΔP_{ECS} and ΔP_{DG} represent regulation

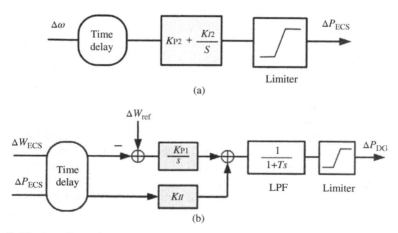

Figure 9.12. Coordinated control loops provided by supervisor agent for: (a) ECS unit and (b) diesel generator.

command signals for the ECS and diesel unit, respectively. (K_{P1}, K_{I1}) and (K_{P2}, K_{I2}) are proportional and integral constant gains for ECS and diesel unit control loops, respectively.

As mentioned, in the proposed secondary control scheme, the ECS provides the main function of MG frequency control and the diesel unit provides a complementary function to support the charging/discharging operation on the ECS unit. Namely, a coordinated control between the ECS and the diesel units has been performed to balance the power demand and the total power generation in the MG [22].

Frequency-dependent battery charging can be used to enhance network frequency regulation capacity. The frequency regulation application could support the power balance related to some renewable energy resources, which are of intermittent nature (e.g., wind and solar powers). As an additional alternative, in coordination with other microsources, the frequency-dependent charging of plug-in vehicles as distributed controllable loads can offer an effective way to improve the system frequency stability. Distributed controllable loads in cooperation with a specified power reserve offer a resource that can rapidly react to the frequency disturbances.

The secondary control also can be used to synchronize the MG before connecting to the main grid, to facilitate the transition from islanded to grid-connected mode. This issue can be usually performed in coordination with MGCC as the global supervisor. Contrary to the local controls, in secondary controls, it may be needed to use low bandwidth communications. Several design examples on MG secondary control are given in Refs [16,17,23–25].

9.5 GLOBAL CONTROLS

Global control deals with some overall responsibilities for an MG, such as interchange power with the main grid and/or other MGs. These controls, which are mainly done by a central controller, are acting in an economically based energy management level between an MG and the neighbors similar to the existing supervisors for power exchanges and economic dispatch in a conventional multiarea power system. To meet the global control objective, wide-area monitoring and estimation is needed for many parameters and indices including fuel and devise storage conditions, commercial power cost and demand charge tariffs, generator reliability, real/reactive power components (power factor), feeder voltages, system frequency, equipment status, predicted weather, current/power spikes, system constraints, and load pattern.

Different control options are investigated for the MG central controller in different MG projects. In the CERTS MG in the United States [26], this controller, called MG energy manager, is responsible for dispatching the output power and the terminal voltage of the DGs. Similarly, in the Hachinohe demonstration project in Japan [27], economic dispatch and weekly operational planning are performed centrally. In the European architecture, it is known as the MG central controller and has several control functions [11].

The MGCC interfaces the MG and the main grid, and also supervises the entire MG units for operations, such as disconnection, reconnection, power flow control, fault level control, market operating, and load shedding. The MGCC may also generate the power output set points for the DGs using gathered local information. Moreover, the MGCC controls power flow at the PCC to keep it close to the scheduled value.

In an MG, identifying the optimal generation schedule to minimize production costs and balancing the demand and supply, which comes from both DGs and the distribution feeder, as well as online assessment of the MG's security and reliability, are the responsibilities of global controls. Global controls supervise the MG's market activities such as buying and selling active and reactive power to the grid and possible network congestions not only in the MG itself, but also by transferring energy to nearby feeders of the distribution network and other MGs. The global controls perform an energy management system (EMS) for MG to ensure a subset of basic functions such as load and weather forecasting, economic scheduling, security assessment, and demand-side management.

The global controls for MG should be implemented through the cooperation of various controllers, located in all other levels, on the basis of communication and collection of information about distributed energy systems and control commands. This could be deployed by optimizing the power exchanged between the MG and the main grid, thus maximizing the local production depending on the market prices and security constraints. This is achieved by issuing control set points to distributed energy resources and controllable loads in order to optimize the local energy production and power exchanges with the main distribution grid [28–30].

Following an islanding event, reconnection of the MG to the main grid can be also done by supervisory control via a controllable switch (SS), and the energy manager (MGCC) sends new power dispatch for participant microsources to provide their proportional share of load in MG. For the grid reconnection, the MG should be synchronized in phase with the main grid, and usually difference in frequency and voltage must be less than 2% and 5% (typically, 0.1 Hz and 3%), respectively. Table 9.1 shows the necessary limit values according to IEEE Standard 1547–2003 [2] for frequency, voltage, and phase angle to achieve a synchronous interconnection between the MG and the main grid.

The local controllers such as MCs and LCs follow the orders of MGCC during grid-connected mode and have autonomy to perform their own controls during islanded mode. Furthermore, the MGCC may have different roles ranging from simple coordination of the local controllers to the main responsibility of optimizing the MG operation [31].

The DNO has the responsibility in managing the operation of medium- and low-voltage grids in which more than one MG may exist. The DNO allows the distribution grid and the connected MGs to operate at an economic optimum and it organizes the

TABLE 9.1. Limits for Synchronous Grid-Connected MG

DG's Average Rating, kVA	Frequency Deviation, Hz	Voltage Deviation, %	Phase Angle Deviation, °
0–500	0.3	10	20
>500–1500	0.2	5	15
>1500–10000	0.1	3	10

relation between the connected MGs and distribution network as well as other connected grids.

As shown in Fig. 9.4, the global control center interfaces the MGCCs of the MGs as well as the distribution network (main grid), and also supervises the power flow control and market operating. This control unit controls power dispatching between the MGs to keep it close to the scheduled values.

In an interconnected MGs network, identifying the optimal generation schedule to minimize production costs and to balance the demand and supply that comes from both MGs and the distribution feeder, as well as online assessment of the MGs security and reliability are the responsibilities of the global control center (market operator). Global control together with the MGCCs supervise the MGs' market activities such as buying and selling active and reactive power to the grid and possible network congestions for transferring energy from an MG to nearby feeders of the distribution network and other MGs. They perform an EMS for the MGs to ensure a subset of basic functions such as load and weather forecasting, economic scheduling, overall security assessment, and demand-side management.

9.6 CENTRAL/EMERGENCY CONTROLS

In an MG, the connected DGs should meet some interconnection standards, and they also must have the capability of intentional disconnection in case of deviating from the specified standards for frequency, voltage, and phase angle (synchronization). For example, based on IEEE Standard 100–2000 [32], operating of DGs with nominal electrical output less than 10 kW in frequency range of 59.3–60.5 Hz is permitted. Otherwise, the DG should be disconnected from the network in no more than 10 cycles (about 0.16 s). For DGs with greater than 10 kW, the operating frequency range is reduced to 59.3–57 Hz.

The voltage constraints for DGs operation in connection mode are also considered by various standards. The requirement for disconnection usually is a function of the voltage deviation. Some cases cite a predetermined number of cycles for disconnection or tripping of DGs for a given voltage range. Typical voltage constraints for under/overvoltage DG trips are given in Table 9.2 [33]. For phase angle constraints, according to the IEEE Standard 2002 [34], typical utility requirements are that the source voltage deviation be no more than +10%, with the source waveform being no more than +10° out of phase with the prevailing utility waveform.

TABLE 9.2. Voltage and Maximum Number of Cycles for Under/Over Voltage DG Trips

Voltage	Maximum Number of Cycles
$V < 50\%$	10
$50\% \leq V < 88\%$	120
$110\% \leq V < 120\%$	60
$V \geq 120\%$	6

In addition to the constraints for the individual microsources, the whole MG should also take advantage of operating in islanding mode, during power outage, blackout, or emergency condition in the main grid, to increase the overall reliability of the power supply. In the emergency condition, an immediate change in the output power control of the MG is required, as it changes from a dispatched power mode to one controlling frequency and voltage of the islanded section of the network. After the initial reaction of the MCs and LCs, which should ensure MG survival following islanding, the MGCC performs the technical and economical optimization of the islanded system.

The islanding plan can be considered as the most important emergency control scheme in the MG systems. When an MG system is islanded, the voltage/frequency might go beyond the power quality limits. Sometimes this transition is likely to cause large mismatches between generation and loads, causing a severe frequency and voltage control problem. Therefore, the islanding procedure requires a careful planning of the existing level of generation and load. In order to ensure system survival following islanding, it is necessary to exploit controllable microsources, storage devices, local load as well as load shedding schemes and special protection plans in a cooperative way [35].

Following islanding, the dependency of frequency and voltage on active and reactive powers allows each microsource to provide its proportional share of load without immediate new power dispatch from the higher level controller, for example, energy manager or MGCC in the global control level. Therefore, in an islanded MG, the small generators are trying to maintain the MG voltage/frequency by controlling the reactive/active power. However, these control actions are not always adequate, and similar to the load shedding in the conventional power systems, following islanding, it may need to curtail some blocks of loads, first from nonsensitive parts.

Therefore, load shedding can be considered as an effective emergency control scheme in the MGs, too. Load shedding can be started in the form of underfrequency or undervoltage load shedding schemes (UFLS, UVLS). The UFLS and UVLS work based on a significant drop in frequency and voltage, respectively. For example, in an islanded situation, when the loads in the MG are higher than total generation capacity, frequency will go down. Therefore, some loads have to be shed to bring the frequency back within the permitted limit.

Similar to the global controls, the emergency controls can be also organized by the MG operator (MGCC). The performance of most existing controls in other levels, as well as the optimal control strategies for the MG depend on the MG's operation state (islanded or grid connected); and switching between control strategies can be done through the operation mode detection. Hence, islanding detection (for unplanned cases) as a significant stage needs more attention, and effective techniques to satisfy the existing standards such as IEEE 1547 [2], IEEE 929–2000 [36], and UL 1741 [37] should be used. The severity of the transients suffered by the MG after an unplanned islanding depends on many factors such as type and place of the disturbance/fault that starts the islanding, operation conditions before islanding, interval until islanding detection, commutation operations subsequent to a disturbance, and type of microsources connected to the MG [7].

Emergency control and protection schemes designed for conventional power systems with unidirectional power flow may become ineffective for modern power systems with numerous distributed MGs and DGs. Undetected faults as well as unnecessary tripping or delayed relay operations may occur due to high DG penetration. It may also disturb the automatic reclosing operation. The operation sequence of protection devices during a fault is thus important [38]. Due to increasing of MGs/DGs, the existing methods used in a fault location could also become inappropriate.

The current operational practice of a distribution network requires the disconnection of MG systems when a fault occurs. This will keep the operational conditions simple and clear, safe and suitable for auto-reclosing. The purpose of MG connection point protection (e.g., frequency and voltage relays) is to eliminate the propagation of fault arcs from the grid to the MG, and to prevent unintended islanding operation.

In an MG, the consequences of an immediate tripping of DG units may become adverse when a sudden change in a power index is seen by other DG units. Even during a fault at an MG network unnecessary disconnection of DG units and microsources may occur due to unwanted trips of feeder or DG unit protection relays, loss of synchronism, sustained overspeed and overcurrent of asynchronous generators or overcurrent and DC overvoltage of power electronic converters. The current operational practice clearly creates a contradiction between network safety and stability.

In the new distribution system with numerous MGs, the protection relays should be used among the gird, on the lowest level like in passive networks. Also, new feeder protection schemes such as directional overcurrent, distance and differential protection, and new fault location applications are needed to be introduced. The protection in MG networks can be improved through advanced protection schemes and decentralized control of DG units.

Using advanced communication/networking technologies as another important issue has a significant role in MGs operation and control. Therefore, the design and implementation of new communication infrastructures and networking technologies for the MGs are key factors to realize robust/intelligent control strategies, specifically in emergency and global control loops. Power line communication (PLC), Internet protocol (IP) based communication network, and wireless networking are common available communication/networking technologies. The employed communication/networking technologies should be capable of supporting the control applications in a secure, efficient, and cost-effective way. On the other hand, the entire network infrastructure in an MG also needs to be controllable and flexible to ensure that every application will perform well and be protected from attack or tampering.

9.7 SUMMARY

The MGs as basic elements of future smart grids have an important role to increase the grid efficiency, reliability, and to satisfy the environmental issues. In this chapter, in addition to the main MG concepts, a comprehensive review on various MG control loops and relevant standards are given with a discussion on the challenges of microgrid controls. Here, all the required control loops in the MGs are classified into primary control, secondary control, global control, and central/emergency control classes.

REFERENCES

1. H. Bevrani, Y. Mitani, and M. Watanabe, Microgrid controls. In: H. Wayne Beaty, editor, *Standard Handbook for Electrical Engineers*, 16th edn, McGraw-Hill, New York, 2013, Section 16.9, pp. 159–176.

2. Standard IEEE 1547-2003, IEEE Standard for Interconnecting Distributed Resources with Electric Power Systems, 2003.

3. B. Kroposki, R. Lasseter, T. Ise, S. Morozumi, S. Papathanassiou, and N. Hatziargyriou, Making microgrids work, *IEEE Power Energy Mag.*, **6**(3), 40–53, 2008.

4. H. Bevrani and T. Hiyama, Automatic generation control (AGC): fundamentals and concepts. In: *Intelligent Automatic Generation Control*, CRC Press, New York, 2011, Chapter 2, pp. 11–36.

5. H. Bevrani, Real power compensation and frequency control. In: *Robust Power System Frequency Control*, Springer, New York, 2009, Chapter 2, pp. 15–38.

6. K. De Brabandere, B. Bolsens, J. Van den Keybus, A. Woyte, J. Driesen, and R. Belmans, A voltage and frequency droop control method for parallel inverters, *IEEE Trans. Power Electron.*, **22**(4), 1107–1115, 2007.

7. A. Llaria, O. Curea, J. Jimenez, and H. Camblong, Survey on microgrids: unplanned islanding and related inverter control techniques, *Renew. Energy*, **36**, 2052–2061, 2011.

8. A. Engler, Applicability of droops in low voltage grids, *Int. J. Distrib. Energy Resour.*, **1**(1), 1–5, 2005.

9. H. Bevrani, Power system control: an overview. In: *Robust Power System Frequency Control*, Springer, New York, 2009, Chapter 1, pp. 1–13.

10. P. Bousseau, F. Fesquet, et al., Solutions for the grid integration of wind farms: a survey, *Wind Energy*, **9**(1–2), 13–25, 2006.

11. B. Awad, J. Wu, and N. Jenkins, Control of distributed generation, *Elektrotech. Informationstechn.*, **125**(12), 409–414, 2008.

12. J. A. Baroudi, V. Dinavahi, and A. M. Knight, A review of power converter topologies for wind generators, *Renew. Energy*, **32**(14), 2369–2385, 2007.

13. B. A. Vaccaro, M. Popov, D. Villacci, and V. Terzija, An integrated framework for smart microgrids modeling, monitoring, control, communication, and verification, *Proc. IEEE*, **99**(1), 119–132, 2011.

14. H. Karimi, E. J. Davison, and R. Iravani, Multivariable servomechanism controller forautonomous operation of a distributed generation unit: design and performance evaluation, *IEEE Trans. Power Syst.*, **25**, 2010.

15. F. Habibi, A. H. Naghshbandy, and H. Bevrani, Robust voltage controller design for an isolated microgrid using Kharitonov's theorem and D-stability concept, *Int. J. Electr. Power*, **44**, 656–665, 2013.

16. F. Habibi, On robust and intelligent frequency control synthesis in the microgrids, M.Sc. thesis, University of Kurdistan, Sanandaj, Iran, 2012.

17. S. Shokoohi, Analysis and control of microgrids under dynamic load variations, M.Sc. thesis, University of Kurdistan, Sanandaj, Iran, 2012.

18. J. A. Peças Lopes, C. L. Moreira, and A. G. Madureira, Defining control strategies for analyzing microgrids islanded operation, *IEEE Trans. Power Syst.*, **21**(2), 916–924, 2006.

19. A. Mohd, E. Ortjohann, D. Morton, and O. Omari, Review of control techniques for inverters parallel operation, *Electr. Power Syst. Res.*, **80**, 1477–1487, 2010.

20. Z. Xiao, J. Wu, and N. Jenkins, An overview of microgrid control, *Intell. Autom. Soft Comput.*, **16**(2), 199–212, 2010.

21. IEEE Standard 1547.3, IEEE Guide for Monitoring, Information Exchange, and Control of Distributed Resources Interconnected with Electric Power Systems, 2007.

22. H. Bevrani and T. Hiyama, Frequency regulation in isolated systems with dispersed power sources. In: *Intelligent Automatic Generation Control*, CRC Press, New York, 2011, Chapter 12, pp. 263–277.

23. H. Bevrani and S. Shokoohi, An intelligent droop control for simultaneous voltage and frequency regulation in islanded microgrids, *IEEE Trans. Smart Grid*, **4**(3), 1505–1513, 2013.

24. H. Bevrani, F. Habibi, P. Babahajyani, M. Watanabe, and Y. Mitani, Intelligent frequency control in an AC microgrid: on-line PSO-based fuzzy tuning approach, *IEEE Trans. Smart Grid*, **3**(4), 1935–1944, 2012.

25. H. Bevrani and T. Hiyama, Neural network based AGC design. In: *Intelligent Automatic Generation Control*, CRC Press, New York, 2011, Chapter 5, pp. 95–122.

26. R. Lasseter, A. Abbas, et al., Integration of distributed energy resources: the CERTS MicroGrid concept, Consortium for Electric Reliability Technology Solutions, California Energy Commission, P50003-089F, 2003.

27. Y. Fujioka, H. Maejima, et al., Regional power grid with renewable energy resources: a demonstrative project in Hachinohe. In: *CIGRE Session, Paris*, 2006.

28. F. Katiraei, R. Iravani, N. Hatziargyriou, and A. Dimeas, Microgrids management, *IEEE Power Energy Mag.*, **6**(3), 54–65, 2008.

29. A. Vaccaro, M. Popov, D. Villacci, and V. Terzija, An integrated framework for smart microgrids modeling, monitoring, control, communication, and verification, *Proc. IEEE*, **99**(1), 119–132, 2011.

30. A. G. Tsikalakis and N. D. Hatziargyriou, Centralized control for optimizing microgrids operation, *IEEE Trans. Energy Convers.*, **23**(1), 241–248, 2008.

31. R. Zamora and A. K. Srivastava, Controls for microgrids with storage: review, challenges, and research needs, *Renew. Sustain. Energy Rev.*, **14**, 2009–2018, 2010.

32. IEEE Standard 100-2000, IEEE Standard Dictionary of Electrical and Electronic Terms, 2000.

33. T. Abdallah, R. Ducey, R. S. Balog, C. A. Feickert, W. Weaver, A. Akhil, and D. Menicucci, Control dynamics of adaptive and scalable power and energy systems for military micro grids, Technical Report ERDC/CERL TR-06-35, Construction Engineering Research Laboratory, 2006.

34. IEEE Standard C62.41.2-2002, IEEE Recommended Practice on Characterization of Surges in Low Voltage (1000 V and Less) AC Power Circuits, 2002.

35. C. C. L. Moreira, Identification and development of microgrids emergency control procedures, Ph.D. thesis, University of Porto, 2008.

36. Standard IEEE 929-2000, IEEE Recommended Practice for Utility Interface of Photovoltaic (PV) Systems, 2000.

37. Standard UL 1741, Inverters, Converters, and Controllers for Use in Independent Power Systems, 2004.

38. P. Jarventausta, S. Repo, A. Rautiainen, and J. Partanen, Smart grid power system control in distributed generation environment, *Annu. Rev. Control*, **34**, 277–286, 2010.

10

MICROGRID CONTROL: SYNTHESIS EXAMPLES

In Chapter 9, the concept of the microgrid (MG) was explained. The MG control loops are divided into four control levels: primary, secondary, global, and central/emergency controls. In this chapter, some design examples for these control levels are briefly discussed. Interested readers can find more details in the authors' previous publications [1–18].

10.1 LOCAL CONTROL SYNTHESIS

10.1.1 Robust Voltage Control Design [4]

As an example for local control design, this section addresses a robust voltage control synthesis technique based on Kharitonov's theorem for an isolated MG. Here, a simple PI structure is used for the voltage controller; however, the PI parameters are tuned by Kharitonov's theorem and the D-stability concept [19]. The proposed PI voltage controller endeavors to minimize errors between direct and quadrature voltage components and their reference values in the presence of parametric uncertainties.

Power System Monitoring and Control, First Edition. Hassan Bevrani, Masayuki Watanabe, and Yasunori Mitani.
© 2014 John Wiley & Sons, Inc. Published 2014 by John Wiley & Sons, Inc.

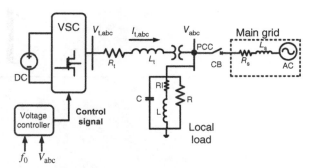

Figure 10.1. An isolated MG system with local controller.

10.1.1.1 Case Study. Schematic diagram of an isolated MG as the case study is illustrated in Fig. 10.1. It contains a DC voltage source, a DC–AC converter interfaced between the DC voltage source and distribution lines, a filter represented by R_t and L_t parameters to extract fundamental frequency of terminal voltage, a three-phase local load depicted by a parallel RLC, a three-phase transformer that transforms voltage from 600 V to 13.8 kV, and a local controller to maintain stability in both connected and disconnected modes. The main grid is described with Rs, Ls, and an AC voltage source. The MG is connected to the main grid via a circuit breaker (CB) at the point of common coupling (PCC) junction. If a disturbance occurs in the power system, such as a short circuit or unit outage, the MG may be unable to maintain its stability in the connected operation. Hence, the circuit breaker will be opened and the MG operating status will be transferred to the islanded mode.

The distributed generator (DG) unit and local loads must be in service in both connected and disconnected operations. In the connected mode, the main grid is responsible for maintaining system voltage and frequency in an acceptable range. In this mode, the voltage source converter (VSC) controls the active and reactive power exchange with the grid using direct-quadrature current control method. As mentioned in Chapter 9, this is known as the *P–Q* control method in which the DG units deliver constant active and reactive powers to the network. In disconnected mode, the MG should be able to control the system voltage and frequency as an independent power grid. This known as the VSI control method in which the inverter controls voltage and frequency in the isolated grid by changing absorbed active and reactive power from DG units.

In an islanded mode, the VSC can employ an internal oscillator with a constant frequency $\omega_0 = 2\pi f_0$ to generate the modulation signals. As shown in Fig. 10.1, the control unit uses rated frequency and three-phase of PCC voltage to control the voltage. For the sake of linear robust control design, the linearized model of the MG is needed. The state space model of the system under balanced condition in abc-frame is given in (10.1), where $V_{t,abc}$, $I_{t,abc}$, V_{abc}, and $i_{L,abc}$ are three-phase terminal voltage, terminal currents, PCC voltages, and PCC currents, respectively.

$$V_{t,abc} = L_t \frac{di_{t,abc}}{dt} + R_t i_{t,abc} + V_{abc}$$

$$I_{t,abc} = \frac{1}{R} V_{abc} + i_{L,abc} + C \frac{dv_{abc}}{dt} \qquad (10.1)$$

$$V_{abc} = L \frac{di_{L,abc}}{dt} + R_1 i_{L,abc}$$

Using α–β transformation and a rotating reference frame, which is presented in Ref. [20], the d- and q-axes of the state space variables system (Fig. 10.1) yield the following equations:

$$\dot{X}(t) = AX(t) + bu(t)$$

$$y(t) = cX(t) \qquad (10.2)$$

$$u(t) = v_{td}$$

where

$$A = \begin{bmatrix} -\dfrac{R_t}{L_t} & \omega_0 & 0 & -\dfrac{1}{L_t} \\[2mm] \omega_0 & -\dfrac{R_t}{L_t} & -2\omega_0 & \dfrac{R_t C \omega_0}{L} - \dfrac{\omega_0}{R} \\[2mm] 0 & \omega_0 & -\dfrac{R_t}{L} & \dfrac{1}{L} - \omega_0^2 C \\[2mm] \dfrac{1}{C} & 0 & -\dfrac{1}{C} & -\dfrac{1}{RC} \end{bmatrix}$$

$$b^T = \begin{bmatrix} \dfrac{1}{L_t} & 0 & 0 & 0 \end{bmatrix}; \quad c = \begin{bmatrix} 0 & 0 & 0 & 1 \end{bmatrix}; \quad X^T = \begin{bmatrix} i_{td} & i_{tq} & i_{Ld} & v_d \end{bmatrix}$$

From (10.2), the transfer function of v_d/v_{td} can be obtained as (10.3). The v_d and v_{td} are direct voltage component of the PCC and inverter terminal voltage, respectively.

$$\frac{v_d}{v_{td}} = \frac{N(s)}{D(s)} \qquad (10.3)$$

where $N(s) = b_2 s^2 + b_1 s + b_0$, and $D(s) = a_4 s^4 + a_3 s^3 + a_2 s^2 + a_1 s + a_0$.

The $N(s)$ and $D(s)$ are numerator and denominator of the transfer function of the open-loop system, respectively. The a_0, a_1, a_2, a_3, b_0, b_1, and b_2 are numerator and denominator coefficients, which are expressed as follows:

$$a_4 = L_t RL^2 C$$

$$a_3 = \left(L_t L^2 + R_t RL^2 C + 2R_t L_t RLC\right)$$

$$a_2 = \left(L_t RL + RL^2 + R_t L^2 + 2R_t R_1 RLC + 2R_1 L_t + R_1^2 L_t RC\right)$$

$$a_1 = \left(R_t RL + 2R_1 RL + R_t R_1^2 RC + R_1 L_t R + 2R_t R_1 L\right.$$

$$\left. + R_t \omega_0^2 RL^2 C + R_1^2 L_t + \omega_0^2 L_t L^2 - 2\omega_0^2 L_t R_1 RLC\right)$$

$$a_0 = R_t R_1 R + \omega_0^2 RL^2 + R_t \omega_0^2 L^2 + R_1^2 R + R_t R_1^2 + \omega_0^2 L_t RL - \omega_0^2 L_t R_1^2 RC - \omega_0^4 L_t RL^2 C$$

$$b_2 = RL^2; b_1 = RL^2\left(\frac{2R_1}{L}\right); b_0 = RL^2\left(\frac{\omega_0^2 L^2 + R_1^2}{L^2}\right)$$

As seen from (10.3), the system has two zeroes and four poles. By substituting the rated values of system parameters given in the Table 10.1, the nominal transfer function of the plant, $g_n(s)$, is obtained as expressed in Equation (10.4). The plant is minimum phase and stable for the nominal values (Table 10.1). However, parametric uncertainties, unpredictable generation variation, load disturbance, and faults may lead the MG to an unstable condition. A robust control may guarantee acceptable performance of the system in this circumstance.

$$g_n(s) = \frac{7.778e7s^2 + 1.101e6s + 2.462e14}{s^4 + 144.2s^3 + 7.789e7s^2 + 2.777e8s + 1.105e13} \tag{10.4}$$

TABLE 10.1. Rated Values for the System Parameters

Quantity	Values
R_t	1.5 mΩ
L_t	300 µH
VSC rated power	2.5 MW
VSC terminal voltage	600 V
PWM carrier frequency	1980 Hz
DC voltage	1500 V
R	76 Ω
L	111.9 mH
C	62.855 µF
Q	1.8
f_0	60 Hz
Transformer voltage ratio	$0.6/13.8\ Kv\ (Yn/\Delta)$

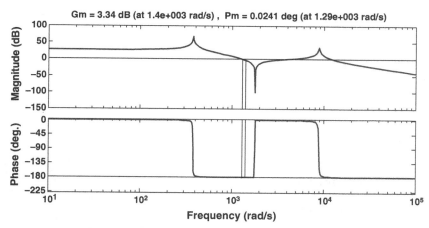

Figure 10.2. Open-loop Bode diagram of the system.

10.1.1.2 Controller Design.

Bode diagrams of the open-loop system are shown in Fig. 10.2. The system gain margin and phase margin are 3.34 dB and 0.0241°, respectively; this indicates a relatively poor stability for the present case study. Based on Kharitonov's theorem [19], a polynomial such as $K(s)$,

$$K(s) = c_0 + c_1 s + c_2 s^2 + c_3 s^3 + c_4 s^4 + \cdots \qquad (10.5)$$

with real coefficients is Hurwitz if and only if the following four extreme polynomials are Hurwitz.

$$
\begin{aligned}
K_1(s) &= c_0^+ + c_1^+ s + c_2^- s^2 + c_3^- s^3 + \cdots \\
K_2(s) &= c_0^- + c_1^- s + c_2^+ s^2 + c_3^+ s^3 + \cdots \\
K_3(s) &= c_0^- + c_1^+ s + c_2^+ s^2 + c_3^- s^3 + \cdots \\
K_4(s) &= c_0^+ + c_1^- s + c_2^- s^2 + c_3^+ s^3 + \cdots
\end{aligned}
\qquad (10.6)
$$

The " − " and " + " show the minimum and maximum bounds on the polynomial coefficients, respectively. In Kharitonov's theorem, the $K(s)$ is a characteristic equation of the closed-loop system.

It is shown [19] that for polynomials with lower than fifth order, there are simpler criteria than (10.6). For a fifth-order plant, it is only sufficient to check the stability of $K_1(s)$, $K_3(s)$, and $K_4(s)$ polynomials. For the problem at hand, the Kharitonov's theorem must be applied on the characteristic equation of the closed-loop system. Considering

(10.3) with PI feedback control, the characteristic equation of the closed-loop system can be obtained as follows:

$$K_{\text{closed-loop}}(s) = s^5 + c_4 s^4 + c_3 s^3 + c_2 s^2 + c_1 s + c_0 \qquad (10.7)$$

where

$$c_4 = a_3$$

$$c_3 = a_2 + b_2 K_p$$

$$c_2 = a_1 + b_2 K_i + b_1 K_p$$

$$c_1 = a_0 + b_1 K_i + b_0 K_p$$

$$c_0 = b_0 K_i$$

Here, $a_i, i = 1, 2, 3$ and $b_j, j = 1, 2$ are coefficients of the open-loop transfer function system and, K_p and K_i are proportional and integral gains of the PI controller, respectively. According to (10.7), the order of the closed-loop system is 5. Therefore, it is just needed to test the Hurwitz criteria for the following three polynomials:

$$K_1(s) = c_0^+ + c_1^+ s + c_2^- s^2 + c_3^- s^3 + c_4^+ s^4 + c_5^+ s^5$$
$$K_3(s) = c_0^- + c_1^+ s + c_2^+ s^2 + c_3^- s^3 + c_4^- s^4 + c_5^+ s^5 \qquad (10.8)$$
$$K_4(s) = c_0^+ + c_1^- s + c_2^- s^2 + c_3^+ s^3 + c_4^+ s^4 + c_5^- s^5$$

It is assumed that with $\pm 10\%$ change in rated values of the system parameters, the coefficients of the open-loop system (10.3) can be perturbed in the following ranges:

$$\left[a_0^+, a_0^- \right] = [1.2155e + 13, \quad 9.9452e + 12]$$

$$\left[a_1^+, a_1^- \right] = [3.0545e + 08, \quad 2.4991e + 08]$$

$$\left[a_2^+, a_2^- \right] = [8.5682e + 07, \quad 7.0103e + 07]$$

$$\left[a_3^+, a_3^- \right] = [1.5863e + 02, \quad 1.2979e + 02]$$

$$\left[b_0^+, b_0^- \right] = [2.7081e + 14, \quad 2.2157e + 14]$$

$$\left[b_1^+, b_1^- \right] = [1.2113e + 06, \quad 9.9108e + 05]$$

$$\left[b_2^+, b_2^- \right] = [8.5559e + 07, \quad 7.0003e + 07]$$

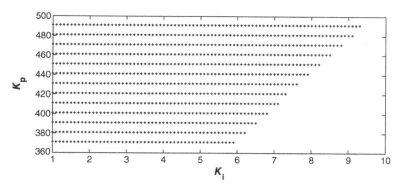

Figure 10.3. Acceptable values of K_p and K_i to stabilize the polynomials given in (10.8).

Following some manual algebraic operations to stabilize (10.8), a set of nine inequalities [4] is obtained. The resulting inequalities are satisfied for some K_p and K_i, which are presented in Fig. 10.3.

In addition to robust stability, to ensure a robust performance such as desirable time response, minimum overshoot/undershoot, and oscillation damping, it is important to maintain the roots of the characteristic equation in a specific region (D region). As seen from Fig. 10.3, there are numerous pairs of K_p and K_i. Using the D-stability concept, proper pairs of K_p and K_i, which remain the roots of the characteristic equation in a specific region are selected. Using this approach, the PI control parameters are tuned as $K_p = 491$ and $K_i = 9.4$.

The basic geometry associated with the zero exclusion condition [19] for $0 < \omega < 50$ kHz is fully demonstrated in Fig. 10.4. The designed PI controller is robust if and only if the rectangle plots do not include the origin. This issue is confirmed in Fig. 10.4. By substituting values of K_p and K_i in (10.7), the characteristic polynomial of the closed-loop system is determined as follows:

$$T(s) = s^5 + 144.2s^4 + 3.827e10s^3 + 1.55e9s^2 + 1.209e17s + 2.314e15 \qquad (10.9)$$

The closed-loop response of the MG system is evaluated against severe step load disturbances, changes in parameters, and operating mode. The simulation results are given in Ref. [4]. It is shown that the proposed control method can handle all test scenarios, effectively.

10.1.2 Intelligent Droop-Based Voltage and Frequency Control [3]

An intelligent method for droop control in an islanded MG based on neuro-fuzzy control technique is presented in Ref. [3]. With an appropriate training, this method can prevent the MG from instability and collapse in the presence of violent changes of load or outage of the distributed generation resource.

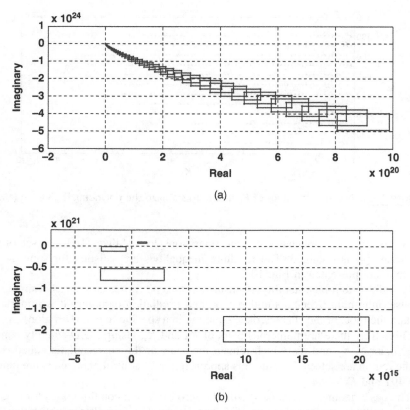

Figure 10.4. The Kharitonov's rectangles: (a) for $0 < \omega < 50$ kHz and (b) a magnified view around the origin.

The neuro-fuzzy control structure is designed to maintain the system stability and minimizing voltage and frequency fluctuations regardless of the MG type and its structure. The most important advantage of the proposed controller is its independence from the MG structure and its operating condition. The participation percentage of active and reactive power in droop-based voltage and frequency controls are conventionally determined by the inductivity and resistivity characteristic of the lines. Based on the resistivity or inductivity of the MG, it is possible to illustrate the effective rate of active and reactive powers in voltage and frequency changes [3]. Since the variation of the load (consumption power) affects voltage and frequency simultaneously, the droop control P/f and Q/V cannot be analyzed, independently.

The main problem of the conventional droop control structure method is in their dependency to the line parameters. Here, an adaptive neuro-fuzzy inference system (ANFIS) is introduced for accurately estimation of these parameters [3].

10.1.2.1 Neuro-Fuzzy Control Framework. The artificial neural network (ANN) is one of the intelligent algorithms that can be used in both identification and control. The ANNs have the ability to learn system behavior and are effectively used for the identification of nonlinear dynamic systems. On the other hand, fuzzy logic (FL) is a powerful tool in control engineering, which can be used to control variable structure systems in the real-time world. The ANNs can be trained by the training data, but the FL has no ability for training. Combining FL and ANNs leads to a useful and valuable result.

Adding the training ability of ANN to FL creates a new hybrid technique, known as ANFIS. In this method, the correction of fuzzy rules is possible when the system is being trained, and by setting the ANN appropriately, previous knowledge about the membership function (MF) and rules is not required, and the optimum MFs are sufficient for obtaining the input/output (I/O) data. The configuration of MFs depends on their parameters. The ANFIS selects these parameters automatically, and does not need a human to obtain these parameters. The parameters of the membership functions are set by means of a back-propagation (BP) algorithm and the least-squares error (LSE) method.

The overall structure of an ANFIS unit is shown in Fig. 10.5 [3], which includes five layers. Weights of the MFs are analyzed in the first layer, which is known as the MF layer. In this layer, the input variables are applied to the MFs. The output of the second layer is the multiplication of the input signals, which are equivalent to IF rules. In the third layer, which is known as the rule base layer, the activity level of even rule is calculated. The number of layers is equal to the number of fuzzy rules. The output of this layer is a normalized form of the previous layer. The fourth layer obtains the output values of the resulting rules. This layer is known as the defuzzification layer. Finally, the output value of the system is obtained from the output layer.

Here, a generalized droop control (GDC) concept [3] is modeled by means of ANFIS, and the validity of the model is examined. Design steps can be summarized as follows:

1. By applying and testing the GDC on the system shown in Fig. 10.6, and then saving the controller inputs/outputs, the training data for the ANFIS controller

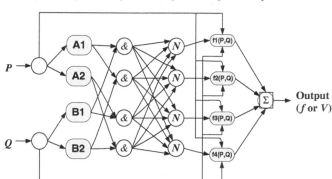

Figure 10.5. Overall structure of an ANFIS.

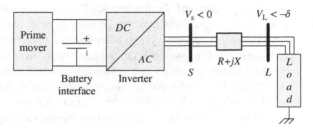

Figure 10.6. A simple MG.

synthesis are collected. To obtain an accurate model, the training data under violent changes of active and reactive loads are considered.

2. After obtaining the training data set, the ANFIS structure is completed. The MFs of input and output are considered in the form of linear and Gaussian functions.

3. After creating the controller structure, using the optimal hybrid method (combination of the LSE and BP), the ANFIS is trained for 5 epochs (iterations) with a small error tolerance (i.e., 0.00001 ms).

10.1.2.2 Synthesis Procedure and Performance Evaluation. A simple voltage source inverter (VSI)-based MG, shown in Fig. 10.6, is used for designing/ training of the ANFIS controller. The ANFIS model is created by using the GDC. The GDC, which is introduced in Ref. [3], consists of two inputs (active and reactive power) and two outputs (voltage and frequency). Since one output is allowed for the construction of ANFIS in the related toolbox in MATLAB software, two ANFIS blocks are used as shown in Fig. 10.7.

Therefore, the considered functions for each block consist of two inputs and one output. After generating the I/O data set and applying them to the *ANFIS* toolbox, the model is trained. Here, to reconstruct the system behavior effectively, the switching frequency of the inverter is fixed at 4000 Hz, and the simulation sampling time is considered as 100,000 samples per second. To evaluate the performance of the designed

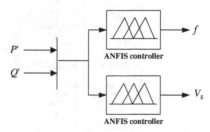

Figure 10.7. The inputs and outputs of the proposed ANFIS controller.

ANFIS model, two sets of data are selected from the real data and training data (as the test data). These two sets are compared and the results are shown in Fig. 10.8. Comparing the real and network data shows whether the training process is done accurately. Figure 10.9 shows the closed-loop control system for a VSI-based inverter.

Two trained ANFIS controllers are replaced with the GDC [3]. After that, under the violent changes of active and reactive loads, the voltage and frequency of the closed-loop system are examined and compared with the generalized droop control structure. The applied scenario for active and reactive power changes is shown in Fig. 10.10a. Figure 10.10b shows the voltage and frequency profiles for both ANFIS and GDC-based methodologies. This figure shows the validity of the ANFIS controller with a high level of accuracy.

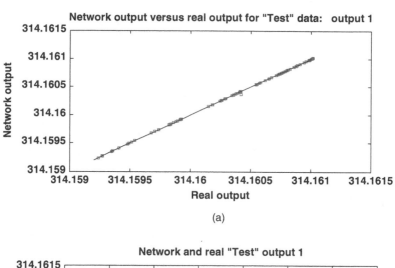

(a)

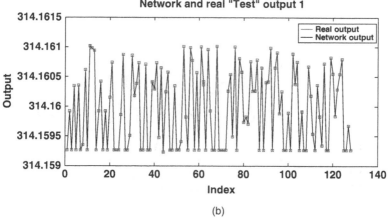

(b)

Figure 10.8. (a) Trained network output versus real output and (b) trained network output and real output together.

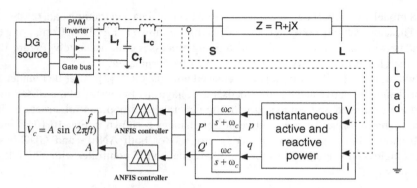

Figure 10.9. A VSI-based MG with ANFIS controller.

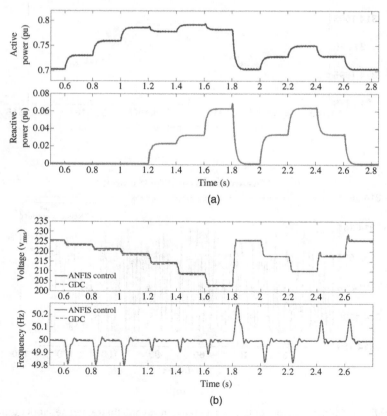

Figure 10.10. (a) Load change pattern scenario and (b) system response.

To prove the reliability of the closed-loop system with the designed ANFIS controller, it is also tested on several MG systems. The results are given in Ref. [3].

10.2 SECONDARY CONTROL SYNTHESIS

10.2.1 Intelligent Frequency Control [2]

The MGs mostly use renewable energy in electrical power production that vary naturally. These changes and usual uncertainties in power systems cause the classic controllers to be unable to provide a proper performance over a wide range of operating conditions. In response to this challenge, an online intelligent approach using a combination of the fuzzy logic and the particle swarm optimization (PSO) techniques for optimal tuning of the most popular existing PI-based frequency controllers in the AC MG systems is addressed in Ref. [2]. The control design methodology is examined on an isolated AC MG case study. The performance of the proposed intelligent control synthesis is compared with the pure fuzzy PI and the conventional Ziegler–Nichols method-based PI control designs.

10.2.1.1 Case Study. An isolated AC MG system is considered as the case study, which is shown in Fig. 10.11. The MG system contains conventional diesel engine generator (DEG), photovoltaic (PV) panel, wind turbine generator (WTG), fuel cell (FC) system, battery energy storage system (BESS), and flywheel energy storage system (FESS). As shown in Fig. 10.11, the DGs are connected to the MG by power electronic

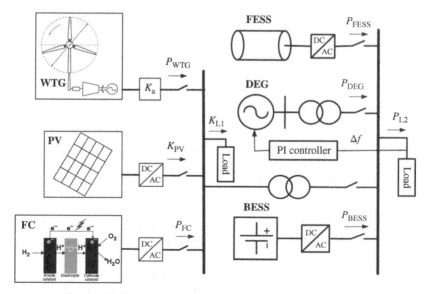

Figure 10.11. Single-line diagram of the AC MG case study.

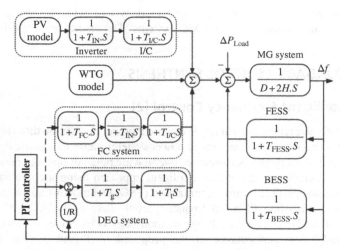

Figure 10.12. Simplified AC MG frequency response model.

interfaces, which are used for synchronization in the AC sources like DEG and WTG and to reverse voltage in the DC sources like PV panel, FC, and energy storage devices.

Each microsource has a circuit breaker to disconnect from the network to avoid the impacts of severe disturbances through the MG or for maintaining purposes. To easily understand the MG frequency response, a simplified frequency response model is given in Fig. 10.12. The FC contains three fuel blocks, an inverter for converting DC to AC voltage, and an interconnection device. Although the FC and some other DGs have high-order characteristic models, for the present study the introduced model (Fig. 10.12) is sufficient.

This model can be useful to analyze/demonstrate the frequency behavior of the case study. The system details including PV and WTG models and the case study parameters are given in Ref. [2]. Since most of the energy sources have an intermittent nature with considerable uncertainty and fluctuation in the power system, efficient control methods must be employed to decrease the undesirable dynamic impacts.

10.2.1.2 *Conventional and Fuzzy PI-Based Frequency Control.* In traditional power systems, the secondary frequency control is mostly done by using conventional PI controllers that are usually tuned based on the specified operating points. In case of any change in the operating condition, the PI controllers cannot provide the assigned desirable performance. On the other hand, if the PI controller is able to track the changes that occur in the power system, the optimum performance will be always achieved. Fuzzy logic can be used as an intelligent method for online tuning of PI controller parameters.

In this section, the traditional PI controller for secondary frequency control is tuned by the well-known Ziegler–Nichols method. Then, a pure fuzzy PI controller is also designed. The result will be compared with the online PSO-fuzzy-based PI design

TABLE 10.2. PI Control Parameters Using the Ziegler–Nichols Method

Controller Parameter	Value
K_p	4.095
K_i	21.84

methodology. A comprehensive study on classical PI/PID tuning methods like Ziegler–Nichols has been presented in Ref. [21]. Using the Ziegler–Nichols method, the PI parameters are obtained as given in Table 10.2.

As described before, to achieve better performance, fuzzy logic is used as an intelligent method. Fuzzy logic is able to respond to the inability of the classic control theory for covering complex systems with their uncertainties and inaccuracies. The control framework for the application of the fuzzy logic system as an intelligent unit in order to fine-tune the traditional PI controller is shown in Fig. 10.13. The fuzzy PI controller has two levels: the first one is a traditional PI controller and the second one is a fuzzy system. As shown, the intelligent fuzzy system unit uses frequency deviation and load perturbation inputs to adjust the PI control parameters. In order to apply the fuzzy logic to the isolated MG system for tuning the PI control parameters, a set of fuzzy rules consisting of 18 rules is used to map input variables, Δf (frequency deviation) and ΔP_L (load perturbation), to output variables, K_p (proportional gain) and K_i (integral gain).

The set of fuzzy rules are given in the Table 10.3 where membership functions corresponding to the input and output variables are arranged as negative large (NL), negative medium (NM), negative small (NS), positive small (PS), positive medium (PM), and positive large (PL). They have been arranged using the triangular membership function, which is the most traditional one. The antecedent parts of each rule are composed by using AND function (with interpretation of minimum). Here, the *Mamdani* fuzzy inference system is also used.

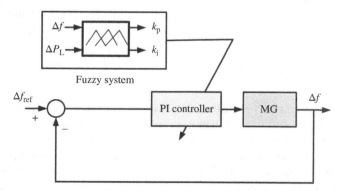

Figure 10.13. Fuzzy PI-based secondary frequency control.

TABLE 10.3. The Fuzzy Rules Set

	Δf					
ΔP_L	NL	NM	NS	PS	PM	PL
S	NL	NM	NS	PS	PS	PM
M	NL	NL	NM	PS	PM	PM
L	NL	NL	NL	PM	PM	PM

It is shown that the fuzzy-based PI performance highly depends on the membership functions [3]. Without precise information about the system, the membership functions cannot be carefully selected, and the designed fuzzy PI controller does not provide optimal performance in a wide range of operating conditions. Therefore, a complementary algorithm is used for online regulation of membership functions.

10.2.1.3 PSO-Fuzzy PI Frequency Control. Here, for online tuning of membership functions employed in the fuzzy PI controller, the PSO is used. The PSO is an optimization algorithm, based on the probability laws, which is inspired by the natural models. This algorithm belongs to a class of direct search methods and is used to find an optimal solution for the optimization problems in a given search space. The applied PSO algorithm is extensively explained in Ref. [2].

Up to now, many search algorithms have been proposed in order to solve the optimization problems, including genetic algorithm, ant colony, and bee colony. However, simplicity is an important advantage of the PSO in comparison with other methods, especially genetic algorithm. Several modifications have been proposed to improve the performance of the PSO algorithm. According to the description provided, what this research investigates is designing an online adaptive controller, using fuzzy logic and PSO, for the purpose of frequency regulation in an AC MG system. The overall control framework for online adjusting of membership functions for the fuzzy rules based on the PSO technique is shown in Fig. 10.14.

Considering the purpose of the algorithm, which is to find the extremum point of the cost function, if the cost function is not properly selected, the algorithm will be stopped in the local extremum points. Initialization of the algorithm parameters is also very important because if they are not carefully selected, the algorithm will never be convergent to the extremum point. The computational flow chart for the proposed online PSO-based optimal design approach is shown in Fig. 10.15.

10.2.1.4 Simulation Results. To compare the classic, fuzzy PI and the PSO-fuzzy PI controllers, several simulation tests are carried out and the performances of the proposed control methods are evaluated. To illustrate the dynamic response of the MG system, the closed-loop system is examined in the face of a multiple step load disturbance, which is plotted in Fig. 10.16a. The MG frequency response using the conventional fuzzy PI and PSO-fuzzy PI (optimal PI) controllers is also shown in Fig. 10.16b.

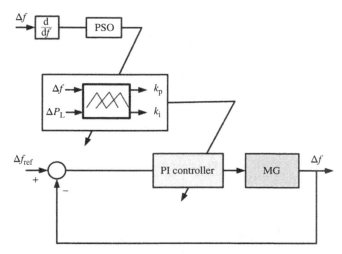

Figure 10.14. Closed-loop system with PSO-fuzzy PI controller.

Δf and ΔP_L are MG frequency deviation and load disturbance, respectively, having values in "pu." As shown, the proposed optimal PSO fuzzy PI controller regulating the system frequency following a disturbance is quite better than the pure fuzzy PI and classical PI controllers. For the sake of a clear comparison between the performance of the PSO-fuzzy PI and fuzzy PI controllers, system frequency following a severe step load disturbance of 0.1 pu is shown in the Fig. 10.17. In this case also, the proposed optimal control method provides a much better performance.

Power system parameters are constantly changing and this may degrade the closed-loop system performance seriously. As indicated in the previous sections, one of the main advantages of the intelligent control methods is robustness against environmental and dynamical changes. To show the adaptive property of the PSO-fuzzy PI controller, the main power system parameters, in the frequency response model (Fig. 10.12), that is, D (damping coefficient), H (inertia constant), R (droop constant), T_t (turbine time constant), T_g (generator time constant), T_{FESS} (FESS time constant), and T_{BESS} (BESS time constant) are significantly changed according to Tables 10.4 and 10.5.

The closed-loop frequency response after applying these changes to the MG system parameters are shown in Figs. 10.18 and 10.19, respectively. It can be seen the conventional controller cannot handle the applied parameters perturbation. Figure 10.19 shows that the difference between the proposed optimal PSO-fuzzy PI controller with the other two controllers is more significant for a higher range of parameter variation.

Finally, two scenarios are examined for the secondary frequency control issue. First, only the DEG is considered as the responsible unit for frequency control in which the results are shown in the previous figures. The impact of the FC contribution in the secondary frequency control is considered as the second scenario. The output of the PSO-fuzzy PI controller is divided between the DEG and the FC units according to their

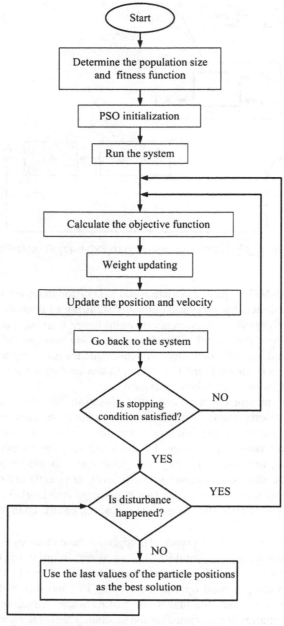

Figure 10.15. The online PSO algorithm flowchart [2].

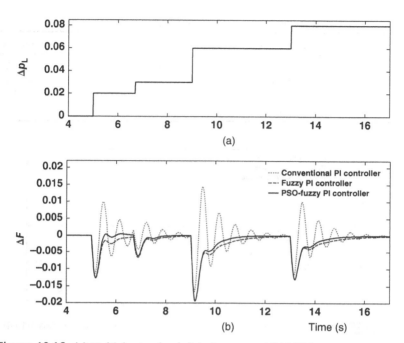

Figure 10.16. (a) Multiple step load disturbances and (b) MG frequency response.

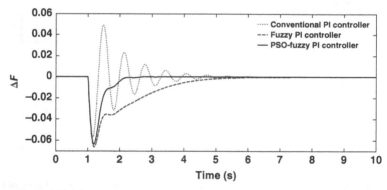

Figure 10.17. Frequency control following 0.1 pu step load disturbance.

TABLE 10.4. Uncertain Parameters and Variation Range

Parameter	Variation Range	Parameter	Variation Range
R	+30%	T_g	+50%
D	−40%	T_{FESS}	−45%
H	+50%	T_{BESS}	+55%
T_t	−50%		

TABLE 10.5. Uncertain Parameters and Variation Range

Parameter	Variation Range	Parameter	Variation Range
R	−60%	T_g	−62%
D	−55%	T_{FESS}	−35%
H	+48%	T_{BESS}	−50%
T_t	−53%		

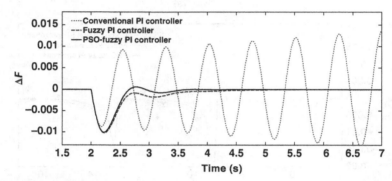

Figure 10.18. Frequency response according to the parameter change (Table 10.4).

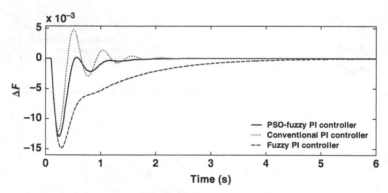

Figure 10.19. Frequency response according to the parameter change (Table 10.5).

participation factors. The result of this cooperation framework using the proposed intelligent technique is shown in Fig. 10.20.

10.2.2 ANN-Based Self-Tuning Frequency Control [5]

Like the conventional generating units, droop control is one of the important control methods for an MG with multiple DG units. The DG units must automatically adjust their

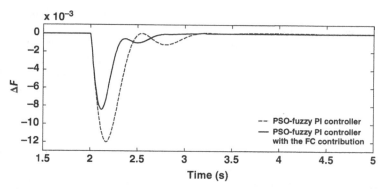

Figure 10.20. Frequency deviation in case of contribution of both FC and DEG in frequency control.

set points using the frequency measurement to meet the overall need of the MG. However, unlike large power systems, the drooping system is poorly regulated in the MGs to support spinning reserve as an ancillary service for secondary frequency control. The main challenge is to coordinate their actions so that they can provide the regulation services.

The possibility of having numerous controllable DG units and MGs in distribution networks requires the use of intelligent, optimal, and hierarchical control schemes that enable an efficient control and management of this kind of systems. Generally, for the sake of control synthesis, nonlinear systems such as MGs are approximated by reduced order dynamic models, possibly linear, that represent the simplified dominant systems' characteristics. However, these models are only valid within specific operating ranges, and a different model may be required in the case of changing operating conditions. On the other hand, due to the increase of nonlinearity and complexity of MG systems, classical and nonflexible control structures may not represent a desirable performance over a wide range of operating conditions. Therefore, more flexible and intelligent approaches are needed [22].

The scheduling of the droop coefficients for frequency regulation in the MGs using an ANN is presented in Ref. [5]. Simulation studies are performed to illustrate the capability of the proposed optimal control approach. The resulting controllers are shown to minimize the effect of disturbances and achieve acceptable frequency regulation in the presence of various load change scenarios.

10.2.2.1 ANN in Intelligent Control.
Neural networks are formed by neurons. An ANN is a crude approximation to parts of the real brain. It is just a parallel computational system consisting of many simple processing elements connected together in a specific way in order to perform a particular task. The ANNs provide a very important tool in optimization tasks because they are extremely powerful computational devices with the capability of parallel processing, learning, generalization, and fault/noise tolerating. Based on configuration and connecting elements, there are several

main applications in ANNs such as brain modeling, financial modeling, time series prediction, control systems, and optimization.

Indeed, an ANN consists of a finite number of interconnected neurons (as described earlier) and acts as a massively parallel distributed processor, inspired from biological neural networks, which can store experimental knowledge and makes it available for use. To use neural networks in optimization tasks, it is needed to have a mathematical model of neural networks. A simplified mathematical model of a neuron is given in (10.10). It consists of three basic components that include weights W_j, threshold (or bias) θ, and a single activation function $f(\cdot)$.

The values W_1, W_2, ..., W_n are weight factors associated with each node to determine the strength of input row vector $X^T = \begin{bmatrix} x_1 & x_2 & \cdots & x_n \end{bmatrix}$. Each input is multiplied by the associated weight of the neuron connection. Depending upon the activation function, if the weight is positive, the resulting signal commonly excites the output node; whereas, for negative weights, it tends to inhibit the output node. The node's internal threshold θ is the magnitude offset that affects the activation of the output node y as follows [22]:

$$y(k) = f\left(\sum_{j=1}^{n} W_j x_j(k) + W_0 \theta \right) \tag{10.10}$$

The neurons could be combined together and they form a layer. Layers are constituted together and make a network. Updating the weights and training of neural networks is based on two basic feed-forward and feedback process. There are three methods for training the weights in feedback process: supervised, unsupervised, and reinforcement learning [22].

In supervised learning, the output is compared with the desired reference vector, and then error vectors applied for updating the weights. There is no desired reference vector for the reinforcement method and a revolutionary process is usually used for updating the weights. In the unsupervised learning method, updating is only based on the input data. In control structures, the supervised learning is usually used. There are several methods for supervised learning such as perceptron, Widrow–Hoff, correlation, and back-propagation learning methods. The most employed method is the back-propagation learning.

The main objective of intelligent control is to implement an autonomous system that could operate with increasing independence from human actions in an uncertain environment. The most common ANN-based intelligent control structures are well explained in Ref. [22]. In all existing structures, the control objectives could be achieved by learning from the environment through a feedback mechanism. The ANN has the capability to implement this kind of learning.

10.2.2.2 Proposed Control Scheme. The schematic diagram of the proposed control scheme for the ANN-based self-tuning frequency controller is shown in Fig. 10.21, where the ANN unit acts as an intelligent unit for optimal tuning of classic PI controller parameters, by getting input and output data based on certain rules. In the

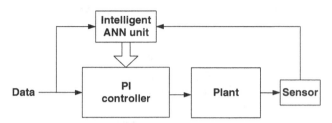

Figure 10.21. Block diagram of the proposed control method.

proposed intelligent control scheme, the ANN collects information about the plant (MG) response, adjusts weights via a learning algorithm, and recommends an appropriate control signal. In Fig. 10.21, the ANN performs an online automatic optimal tuner for the existing PI controller. The main components of the ANN as a fine-tuner for the PI controllers include a response recognition unit to monitor the controlled response and extract knowledge about the performance of the current controller gain setting, and an embedded unit to suggest suitable changes to be made to the controller gains.

The employed neural network structure for tuning the parameters of the PI controller used for frequency control (of the system given in Fig. 10.11) is shown in Fig. 10.22. In the ANN structure of Fig. 10.22, 20 linear neurons are considered for network input layer; 10 and 2 nonlinear neurons are also considered for hidden layer and output layer, respectively. The number of output layer's neurons is equal to the number of control parameters that must be adjusted. In Fig. 10.22, X is the input vector and W_1 and W_2 are connecting weight vectors between the layers.

Selection of initial conditions in an ANN-based control system is also known as an important issue. In multiobjective control problems, some initial values may not guarantee the achievement of objectives with a satisfactory value of the optimization function. The initial conditions are usually selected according to the *a priori* information about distributions at the already-known structure of the open-loop ANN and selected

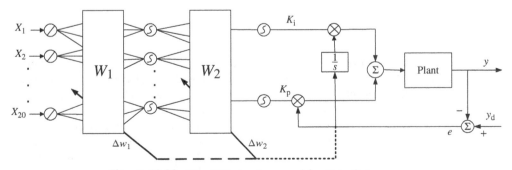

Figure 10.22. The ANN structure used for PI tuning.

control strategy. Here, the initial quantities in the applied ANN scheme (Fig. 10.22) are considered as follows:

$$X = \text{ones}(20, 1)$$
$$W_1 = \text{rand}(10, 20)$$
$$W_2 = \text{rand}(2, 10)$$

As mentioned, linear functions are considered for the first layer, and for the second and third layers, the sigmoid functions are chosen. In the output layer, different coefficients for sigmoid function are considered for tuning of the controller parameters. The main advantage of using these nonlinear functions is in performing a smooth updating of weights.

The learning process of the applied ANN for the MG test system is to minimize the performance function given by (10.11), where y_d represents reference signal and y represents the output of output layer.

$$E = \frac{1}{2}(y_d - y)^2 \qquad (10.11)$$

The implemented algorithm for updating weights is based on back-propagation learning, which is described in the flowchart of Fig. 10.23. In the feed-forward process, by using the input vector (X), the values of hidden layer output (H) and output layer result (O) are provided, and then error value (E) obtained from the process is employed to update the weights as follows:

$$w_2(k + 1) = w_2(k) + \Delta w_2 = w_2(k) + \eta \delta H$$
$$w_1(k + 1) = w_1(k) + \Delta w_1 = w_1(k) + \eta \sigma X \qquad (10.12)$$

where η is a *learning rate* given by a small positive constant, and

$$\sigma \equiv -\frac{\partial E}{\partial y} \qquad (10.13)$$

$$\delta = (y_d - y)f'(H) \qquad (10.14)$$

The learning process continues to reach the desired minimum error. This method is presented in detail in Ref. [22]. For testing of the proposed control methodology for tuning the PI controller parameters, the controller is applied to the case study (Fig. 10.11), and the results are compared with the response of a conventional PI controller.

To examine the proposed control strategy, system frequency response in the presence of the ANN-based self-tuning PI controller and conventional PI controller is tested. Here, a random step load disturbance, WTG mechanical output power variations, and sunlight flux variations are simultaneously considered in the MG test system

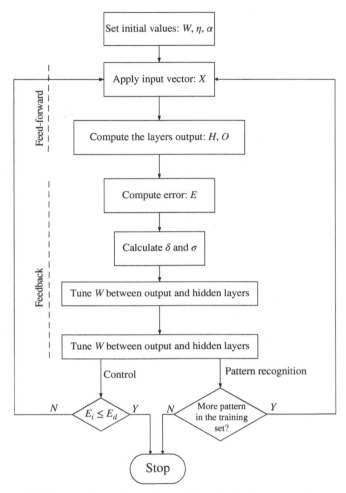

Figure 10.23. Flowchart of updating weights via back-propagation learning.

to better evaluate the proposed closed-loop control performance. In this scenario, it is assumed that the BESS and FESS systems do not participate in the secondary frequency control issue. The results for the first scenario are plotted in Fig. 10.24a–e.

The considered power fluctuations, the WTG and PV output power variations, the DEG frequency response, and system frequency are shown in Fig. 10.24a–e, respectively. As shown, only the DEG participates in the secondary frequency control process and it injects more compensating power by using the proposed intelligent control method. Therefore, when the ANN adjusts the controller parameters, system frequency fluctuations are much less.

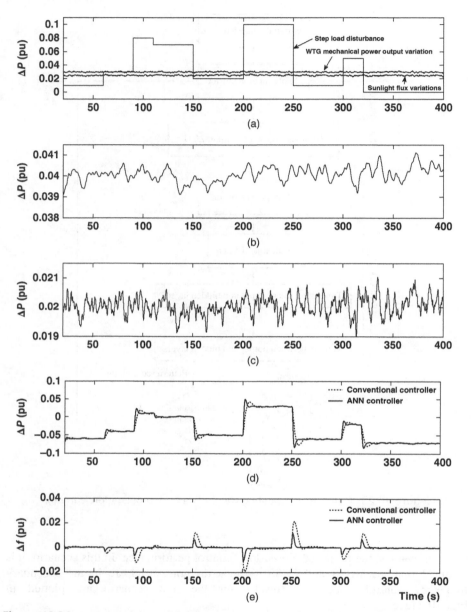

Figure 10.24. System response: (a) step load disturbances, and WTG and PV power, (b) WTG power, (c) PV power, (d) DEG power, and (e) system frequency.

10.3 GLOBAL CONTROL SYNTHESIS

As a considerable capability, each MG can operate in autonomous (isolated from the main grid) and grid-connected modes. The performance measure in the autonomous mode is the reliability of stand-alone operation. However, in the grid-connected mode the MG operates while connected to the main grid. This is especially characterized by the fact that each MG can sell a portion of its generated power to the grid at a point of connection and at the same time is able to purchase a portion of its demand from the grid at another point of connection. As a result of power sharing in this mode, load demand supplement is guaranteed at all times by the grid.

The increasing amount of demand for electrical energy along with growing environmental concerns motivate the idea of establishing new power systems with flexible and intelligent programs of generation as well as demand-side management. These programs run by utility companies aim to provide consumers with a reliable and cost-efficient energy and at the same time to make efficient use of the generation and transmission infrastructure. While many of these programs are still under investigation, there already exist a number of practical applications in many countries across the world.

In this direction, energy consumption scheduling and power dispatching can be considered as important global control issues in distribution networks with interconnected MGs

10.3.1 Adaptive Energy Consumption Scheduling [7]

Demand side can be managed by either reducing or shifting the consumption of energy. While the former can be efficient to some extent, the latter proposes shifting of high load household consumptions to off-peak hours in order to reduce *peak-to-average ratio* (PAR). The high PAR might lead to degradation of power quality, voltage problems, and even potential damages to utility and consumer equipment.

With the advancement of smart metering technologies and increasing interest in power distribution networks with two-way communications capability, load management has appeared in the form of energy consumption scheduling (ECS). In the ECS, the power consumption time of connected units is optimally scheduled so that some interesting measures such as generation cost and PAR can be optimized efficiently. This results in reducing the risk of getting into a condition that may lead to a blackout. As an incentive that subscribers follow ECS decisions, intelligent pricing schemes in the form of lower utility charges should be provided.

Consequently, customers will be encouraged to shift their heavy loads to off-peak hours. These issues motivate the design of ECS with the aim of minimizing power generation cost and PAR. The proposed ECS schemes in the literature mainly perform network-wide load management with the assumption of the knowledge of the whole network demand *a priori* or at least with known statistical characteristics. In other words, a network operator should be aware of the whole network demand in some way. Due to the diversity of power customers ranging from household to industrial domains with uncertain demands, however, this case is not *mostly* valid. Alternatively, an operator who is aware of demand in a local area, not other neighbor areas, might be interested in ECS within this area. The fact that aggregate power generation cost depends on the network-wide demand necessitates considering the impact of uncertain demands in the design of the ECS.

To investigate the mentioned difficulty, a distribution network connecting to a local area (LA) consisting of several MGs with known demand in average and other neighbor areas (NAs) with uncertain demands is considered. The network operator performs ECS of demand in the LA considering NAs demand as a random variable. This ECS is formulated with two stochastic optimization problems, one with the objective of the network-wide power generation cost minimization and the other with the objective of PAR minimization. While these two objectives are correlated to some extent, optimizing one does not necessarily imply the optimality of the other.

These objectives are compared using optimal, adaptive, and uniform scheduling schemes in terms of generation cost and PAR. In the optimal one, the optimal solutions of two underlying problems are achieved with the assumption of the knowledge of NAs demand in advance. Without this assumption for practical purposes, an adaptive scheme with *online* stochastic iterations to capture the randomness of uncertain demands over the time horizon continually is proposed. Finally, in uniform scheduling, the demand of MGs in LA is uniformly distributed over the time horizon regardless of NAs demand.

10.3.1.1 Distribution Network with Connected MGs. A simplified architecture of a distribution network organized by a distribution company (Disco1) is shown in Fig. 10.25. This network consists of N connected MGs in an LA, and NAs that may belong to another company (Discoi). The microsources and storage devices use power

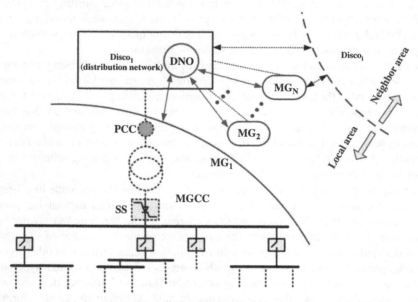

Figure 10.25. A distribution network with connected MGs.

electronic circuits to connect to the MG. The MG can be connected to the network by a PCC via a static switch. This switch is capable of islanding the MG for maintenance purposes or when faults or a contingency occurs.

The DNO deals with some overall responsibilities for the distribution network (Disco) and the connected MGs, such as interchange power between the main grid and the MGs. This unit, which is located in the application layer of the distribution management system is acting in an economically based energy management between the main grid and the neighboring MGs.

As shown in Fig. 10.25, the DNO interfaces the main grid (Disco1) with the connected MGs (in the LA) as well as other neighbor grids (in NAs that may be covered by another Disco). The DNO also supervises the power flow control and market operation. This operator controls power flow from the main grid to the MGs to keep it close to the scheduled values. In the mentioned network, identifying the optimal consumption/generation schedule to minimize production costs and to balance the demand and supply, as well as online assessment of security and reliability, are the responsibilities of the DNO unit. As mentioned in Chapter 9, the DNO together with the MGCCs supervise the MG's market activities such as buying and selling active and reactive power to the grid and possible network congestions for transferring energy from a distribution network to the MGs in a local area and other neighboring areas.

10.3.1.2 Proposed Methodology and Results. Pricing of electricity can be used as a mechanism to encourage customers to follow a specified load scheduling. Various pricing schemes have been proposed by economists and regulatory agencies such as flat pricing, critical-peak pricing, time-of-use pricing, and real-time pricing. Among them, real-time pricing is motivated to be used in the next-generation power systems concerning its environmental and economic gains. Accordingly, in the present work, an energy scheduling approach based on real-time generation cost is proposed, which can be used to establish a real-time pricing scheme.

In the mentioned network, the objective of DNO to implement ECS could be either to minimize power generation cost or to minimize PAR. A solution is summarized as an adaptive cost-aware ECS (ACA-ECS) algorithm [7]. In order to minimize PAR of the total instantaneous power delivered to LA and NAs during the time period T, a minmax formulation is proposed with the objective of minimizing the peak of this power. The mathematical formulations are given in Ref. [7].

For investigation of the proposed methods, a distribution network is considered in connection with an LA consisting of $N = 10$ MGs and an NA. The ECS located in the DNO schedules energy consumption of MGs in LA during a time horizon of length 6 h.

The impact of the proposed ACA–ECS algorithm on the time domain curvature of the total grid demand in the presence of an *unknown* NA demand is evaluated and the result is compared with an optimal technique. A typical realization of $T = 360$ samples of NA demand with standard deviation of $\sigma = 20$ kWh and the corresponding optimal total demand are shown in Fig. 10.26a. As observed, the optimal solution schedules LA demand such that the system-wide total demand becomes smooth suitable for cost minimization. In fact, scheduling the LA demand provides a diversity for the ECS to mitigate the stochastic nature of NA demand. Total demand using the ACA–ECS scheme

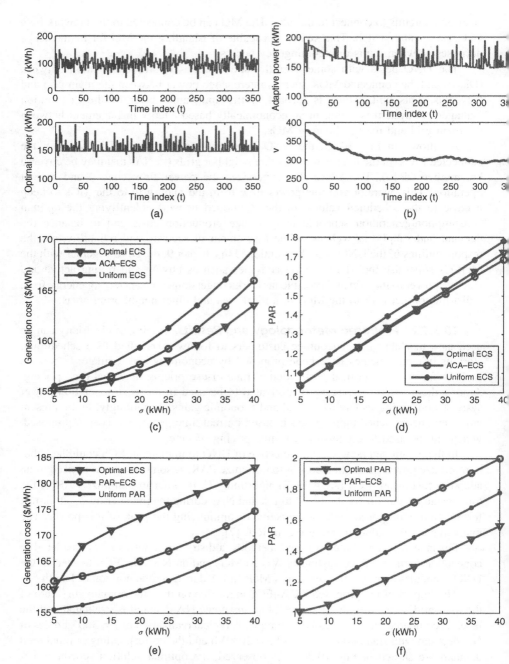

Figure 10.26. Simulation results: (a) generated NAs demand and the system-wide optimal demand, (b) adaptive system-wide demand and Lagrange multipliers, (c) generation cost per kWh in cost formulation, (d) PAR in cost formulation, (e) generation cost per kWh in the PAR formulation, and (f) PAR in PAR formulation.

and the corresponding Lagrange multipliers [7] are also shown in Fig. 10.26b. Intuitively, after some initial time slots, the behavior of the total demand curve approximately converges to that of the optimal solution in Fig. 10.26a.

Performance measures of the ACA–ECS and the optimal ECS schemes, such as generation cost per kWh and PAR, versus the randomness of the NA demand could be interesting in the following. As another scheduling scheme, the results of uniform ECS scheme are also included. In this scheme, the demand of each MGn in LA is *uniformly* distributed over the whole time horizon, independent of the NA demand. This can also be considered as a deterministic solution. Cost and PAR performances versus the standard deviation are shown in Fig. 10.26c and d, respectively. For each instance, similar to the time-domain performance, first a data set with $T = 360$ samples are generated. This set is used to obtain the optimal ECS solution once in the beginning of the time horizon as well as to provide the ACA–ECS scheme with instantaneous realized NA demand. As shown in the first part of Fig. 10.26a, there is a typical generated data set with $\sigma = 20$ kWh. As a common observation in Fig. 10.26c and d, performance measures get worse as σ increases. In the case of cost measure, this is due to the fact that the considered squared cost function results in higher cost per kWh for high demand values in comparison with low demand values. The results in PAR are based on the fact that the averages of both LA and NA demands are made constant when σ increases. Considering PAR as a fractional term of the peak demand over the average demand, it is reasonable to conclude that PAR increases as σ increases. Moreover in Fig. 10.26c, with increase in σ, the performance gap between the compared ECS schemes and the optimal one increases. In the case of ACA–ECS, this is due to the fact that the stochastic estimator (given in Ref. [7]) would be far from optimality with the increase in the randomness of the uncertain part of power generation. In the case of uniform ECS, the degradation effect of high randomness would be more severe since this scheme does not take care of NA demand in the scheduling decisions.

Furthermore, in Fig. 10.26c and d, the generation cost and PAR performances of the ACA–ECS scheme outperform those of uniform ECS. This is reasonably expected as ACA–ECS takes advantage of the diversity in NA demand to smooth the total demand and therefore achieves a better performance. In the comparison between ACA–ECS and the optimal solution, it is observed that the optimal solution achieves lower cost. This is due to the fact that this solution fully takes into account the knowledge of NA demand at the beginning of the time horizon for the scheduling of LA demand. However, ACA–ECS makes a scheduling decision adaptively per a time unit, when the demand of NA is available in that unit. Remarkably, the PAR of ACA–ECS is comparable to that of the optimal solution. This implies that the optimality of generation cost does not necessarily imply the optimality of PAR too. This observation motivates the performance evaluation of PAR formulation in the following.

In order to evaluate the efficiency of PAR formulation, the generation cost per kWh and PAR performances of this formulation are illustrated in Fig. 10.26e and f, respectively. Similar to the cost formulation [7], the results of optimal solution in PAR formulation (optimal PAR) and uniform scheduling (uniform PAR) scheme are also included. Since the scheduling of the uniform strategy is independent of the objective function, the achieved results are the same in both cost and PAR formulations. We take advantage of this equality and take uniform strategy curves as references for comparison between these formulations.

Comparing Fig. 10.26c and e, it is observed that uniform scheduling was the worst in the former, whereas it is the best in the latter. Considering the results of uniform scheduling as reference in both figures, we conclude that cost minimization formulation is more cost efficient in comparison with PAR formulation. In terms of PAR, the optimal solution in the PAR formulation achieves the lowest PAR. This is reasonably expected as this solution takes NA demand into account *a priori*. In comparison with the uniform strategy, the PAR of PAR–ECS scheme is high. More importantly, this implies that PAR performance of the proposed adaptive approach in cost minimization formulation even outperforms its equivalent in the PAR minimization formulation.

This observation along with the lower generation cost in cost formulation demonstrates that our proposed adaptive approach achieves more *efficient* results with this formulation compared with the PAR one. Also, the proposed adaptive approach is a trade-off between the optimal (full NAs demand) and uniform (no NAs demand) schemes in terms of generation cost and PAR minimization.

10.3.2 Power Dispatching in Interconnected MGs [8]

Load demand management is a critical issue in the smart grids with power sharing capability. It controls the power dispatching between MGs with the aim of establishing a balance between power supply and demand in a cost-efficient manner. The objective in this balance is to alleviate peak loads at individual MGs and accordingly avoids major expenditures in power utilities. In contrast to the autonomous mode where load management results in shifting peak loads to off-peak loads, peak loads of an MG in grid-connected mode can be handled by the means of power sharing throughout the grid.

Considering stochastic demands and maximum allowed supplies in MGs, an immediate question is how to perform power dispatching and to set interactions in the grid. Due to power generation and transmission costs, this issue raises the economic exploitation of the resources within the grid. The outcome of power dispatching could be interesting in this perspective.

In Ref. [8], load demand management of an electric network of interconnected MGs is formulated as a power dispatch optimization problem. Real-time pricing is employed as a motivation for interactions between the MGs. The objective is to minimize the network operational cost and at the same time to satisfy the stochastic demands within the MGs in average. With the solution of this problem, a cooperative power dispatching algorithm between MGs is proposed under the assumption of a communication infrastructure within the grid. The core parameter in this algorithm is a defined dynamic *purchase price* per unit of power at each MG. Considering their demands and supplies, the MGs progressively update and broadcast their prices throughout the grid. Every MG adaptively regulates its transactions with the rest of the grid by taking into account its realized demand as well as already announced prices from the other MGs. This strategy results in a semidistributed and reliable load management within the grid in a cost-efficient manner.

10.3.2.1 *Methodology and Results.* As an example, consider a power grid consisting of small MGs starting from MG1 and ending with MG8 as shown in Ref. [8]. The distance between any two neighbor MGs is the same and is noted as one hop. The transmission price per unit of power between any two MGs is assumed to be the number of hops between them. Moreover, the power generation function is assumed to be a square function.

The simulation is run for 200 time slots; each slot with realization of a new demand and new maximum permitted supply values per MG. Produced powers at individual MGs by the proposed statistical cooperative power dispatching (SCPD) algorithm are shown in Fig. 10.27a. The s_n in this figure represents the produced power of MG_n. Despite the diverse power demands, the produced powers are close to each other. In comparison with their demands, low demand MGs produce higher power and high demand ones produce lower power. This is due to the fact that low demand MGs tend to sell power, whereas high demand ones tend to purchase power. Figure 10.27b illustrates Lagrange multipliers λ_i by which MGs interact within the grid. As previously mentioned, each λ_i in the SCPD algorithm can be interpreted as the price that announces at time slot to pay for a unit of

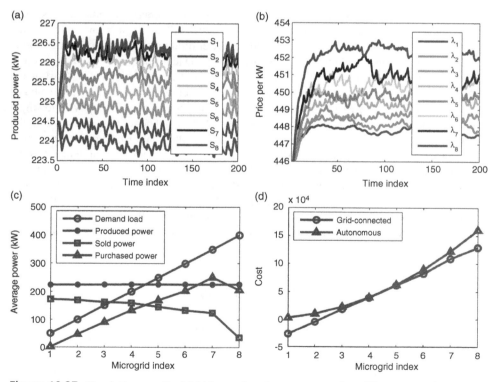

Figure 10.27. Simulation results: (a) MGs produced powers over time, (b) announced power prices over time, (c) average load demands, and produced, sold and purchased powers, and (d) cost comparison between grid-connected SCPD and autonomous modes.

power from any other MG in the system. As shown, even though all prices are initialized with the same value, they statistically converge to different levels during the simulation. The higher is the demand, the higher is the announced price for power purchasing.

The curves in this figure are interpreted relative to each other, that is, power flow direction within the grid is from MGs with low prices to ones with high prices. In other words, an MG with low purchase price in comparison with others is an indication of power selling and vice-versa.

The average *demand, produced, sold*, and *purchased* powers of individual MGs in the SCPD algorithm are shown in Fig. 10.27c. While demand increases linearly with MG indexes in accordance with assumed values, produced powers are approximately the same for all MGs, in compliance with Fig. 10.27a. Remarkably, the sum of produced and purchased powers at each MG is equal to the sum of demand plus sold power. This reveals the energy balance within the grid. Purchased and sold power curves vary in opposite directions versus the MG index, that is, purchased power increases with demand, whereas sold power decreases. Exceptionally, the decrease of purchased power is possibly due to its location in the network topology, which burdens high power transmission cost. In a logical statement, low demand MGs sell power to high demand ones.

This statement is investigated by comparing the average operational costs at individual MGs in these two modes, shown in Fig. 10.27d. To this end, the cost in autonomous mode is obtained from the square power production function. Furthermore, the cost in grid-connected mode implemented by SCPD algorithm is the production cost plus purchased cost minus the revenue from selling power to the other MGs. As shown, the grid-connected mode achieves lower cost for low demand and high demand MGs. The decrease in the cost of low demand MGs is the result of selling power to high demand ones. In particular, the cost of MG_1 and MG_2 even get negative as a result of high revenue from selling power that compensates their production cost. This outcome also decreases the cost of high demand MGs as they purchase a portion of their demand from the low demand ones. This is accomplished by means of high announced purchase prices in Fig. 10.27b. Furthermore, the purchased and sold power of moderate demand MGs are mostly the same. In summary, as a numerical indicator, the proposed SCPD in the grid connected mode achieves 20% cost reduction in comparison with stand-alone operation. Overall, this power sharing scheme transforms the parabolic cost curve to a linear one as shown in Fig. 10.27d.

10.4 EMERGENCY CONTROL SYNTHESIS

Like conventional systems, having acceptable voltage and frequency levels is necessary in an MG. In grid-connected mode, this issue is realized by the main grid because distributed generators are often based on active and reactive power $(P-Q)$ controlled inverters with specific amount of active and reactive powers. However, in islanded mode, the MG security strongly depends on the capability of existing MG controllers in primary, secondary, and emergency levels. Also, power exchange between the main grid and MG in a grid-connected mode leads to voltage and frequency variations during transient between two operation modes. Stable performance of the MG in the transient and islanding states is dependent on the VSI-based DGs.

As mentioned before, after disconnection the control mode of one or several DGs will change from the $P–Q$ control to the VSI control. However, this control may be unable to stabilize an islanded MG following large disturbances like load disturbance and generator trip, which change the balance between generation and consumption. In such cases, emergency control may need to be activated. In this section, an effective load shedding (LS) scheme based on both voltage and frequency records as the last action to prevent the system blackout is suggested.

10.4.1 Developed LS Algorithm [10,11]

The proposed LS algorithm is summarized in a flowchart, which is shown in Fig. 10.28. The LS scheme uses both frequency and voltage measurements as well as the rate of change of frequency (ROCOF). Here, the amount of voltage decrease following a disturbance is used to determine the number of load blocks to be shed; therefore, voltage parameter is measured at first. Any significant change in voltage causes the algorithm to be started. However, to prevent unnecessary LS, frequency and ROCOF are also measured during the specified recording time window (e.g., τ s). If they pass the determined thresholds, the LS will be initiated.

Amount of frequency and ROCOF thresholds can be defined based on a fault condition in which only one load block is required to be shed. On the other hand, the amount of voltage reduction determines the severity of disturbance/fault. The number of load blocks for the LS is calculated from a look-up table. This table is determined based on voltage deviation range following disturbances. Recording of voltage drop for different events and performing of a look-up table can be achieved by increasing loads until the system is stabilized by eliminating all needful load blocks for the LS.

When the LS is initiated, the removal of the first group of load blocks (N) will be done from the look-up table. However, it may need to shed more blocks following a delay due to breakers operation. After T seconds, the frequency is measured again to ensure that it is not reaching its nominal value. This process will be continued until the frequency returns to the nominal value.

10.4.2 Case Study and Simulation

Here, a simple MG which is used in [23], is considered as the case study. The base voltage of the MG is 13.8 kV, which is connected to a 69 kV distribution grid through a substation transformer and breaker. The system includes four feeders and three DGs. Two electronically interfaced DGs (DG_2 and DG_3) and one conventional unit (DG_1) are located in feeders 3, 4, and 1, respectively. It is assumed that the available loads are divided into critical and noncritical RL loads. The noncritical loads are shed in the necessary conditions.

The proposed LS plan is applied to the case study, following loss of DG_2 at $t = 0.7$ s. Simulation results show that after this disturbance and islanding, the MG system

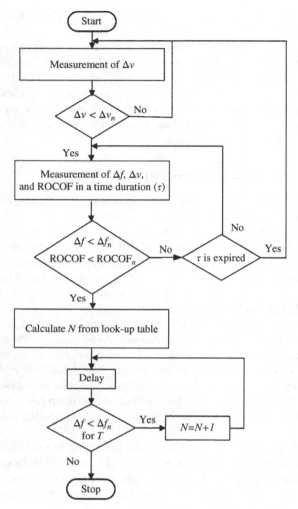

Figure 10.28. The LS algorithm.

becomes unstable. Therefore, the LS is required to stabilize the islanded MG system. Here, based on dynamical knowledge obtained mainly via simulations, the look-up table (Table 10.6) is performed and the required thresholds are determined.

Following this disturbance, the voltage decreases to 0.774 pu. According to the look-up table, the number of load blocks (N) for LS is 1. Therefore, one block is shed at $t = 0.886$ s. After T and delay time, the frequency goes back to the normal range. This means that it is not necessary to shed more load blocks. The system response due to the use of the proposed LS is presented in Fig. 10.29.

In an islanded MG that has reached equilibrium after islanding, the voltage of all buses and branches changes following a disturbance. If a generation unit is tripped all

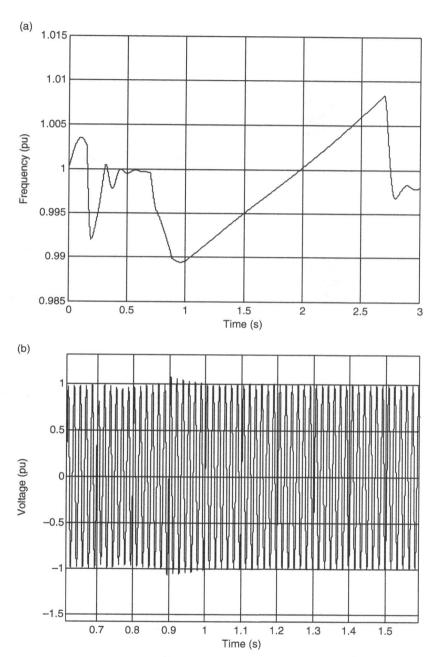

Figure 10.29. System response after using the LS scheme: (a) MG frequency and (b) bus voltages.

TABLE 10.6. Look-Up Table of Load Shedding

Voltage Deviation Range, pu	Number of Load Blocks for LS (N)
$V \geq 0.776$	0
$0.776 < V \leq 0.733$	1
$0.733 < V \leq 0.71$	2
$0.71 < V \leq 0.696$	3
$0.696 < V \leq 0.68$	4
$0.68 < V \leq 0.67$	5
$V \leq 0.67$	All load blocks for load shedding

branch voltages will decrease, except the related branch of the removed generator. At the moment of separation, voltage owned the branch of the tripped unit increases and remains constant at this value. This voltage increase in any branch of generation unit is the sign of loss of that unit.

It is noteworthy that an LS is impressionable by the location of shed loads. However, whenever the amount of load blocks to be shed is more the effect of location is less. In the Fig. 10.29, the load with first priority was shed.

10.5 SUMMARY

This chapter addresses several synthesis methodology examples for controller design in an MG. These examples cover all control levels, that is, primary, secondary, global, and emergency controls. The applied algorithms and control techniques are mostly based on robust, intelligent, and optimal/adaptive strategies.

REFERENCES

1. H. Bevrani and T. Hiyama, Frequency regulation in isolated systems with dispersed power sources. In: *Intelligent Automatic Generation Control*, CRC Press, New York, 2011, Chapter 12, pp. 263–277.

2. H. Bevrani, F. Habibi, P. Babahajyani, M. Watanabe, and Y. Mitani, Intelligent frequency control in an AC microgrid: on-line PSO-based fuzzy tuning approach, *IEEE Trans. Smart Grid*, 3(4), 1935–1944, 2012.

3. H. Bevrani and S. Shokoohi, An intelligent droop control for simultaneous voltage and frequency regulation in islanded microgrids, *IEEE Trans. Smart Grid*, 4(3), 1505–1513, 2013.

4. F. Habibi, A. H. Naghshbandy, and H. Bevrani, Robust voltage controller design for an isolated microgrid using Kharitonov's theorem and D-stability concept, *Int. J. Electr. Power*, 44, 656–665, 2013.

5. H. Bevrani, F. Habibi, and S. Shokoohi, ANN-based self-tuning frequency control design for an isolated microgrid. In: P. Vasant, editor, *Meta-Heuristics Optimization Algorithms in Engineering, Business, Economics, and Finance*, IGI Global, Hershey, PA, 2012, Chapter 12, pp. 357–385.

6. P. Babahajyai, F. Habibi, and H. Bevrani, An on-line PSO-based fuzzy logic tuning approach: microgrid frequency control case study. In: P. Vasant, editor, *Handbook of Research on Novel Soft Computing Intelligent Algorithms: Theory and Practical Applications*, IGI Global, Hershey, PA, 2014, Chapter 20, pp. 589–616.

7. M. Fathi and H. Bevrani, Adaptive energy consumption scheduling for connected microgrids under demand uncertainty, *IEEE Trans. Power Deliv.*, **28**(3), 1576–1583, 2013.

8. M. Fathi and H. Bevrani, Statistical cooperative power dispatching in interconnected microgrids, *IEEE Trans. Sustain. Energy*, **4**(3), 586–593, 2013.

9. H. Bevrani, M. Gholami, and N. Hajimohammadi, Microgrid emergency control and protection: key issues and new perspectives, *Int. J. Energy Optim. Eng.*, **2**(1), 78–100, 2013.

10. N. Hajimohammadi and H. Bevrani, Load shedding in microgrids. In: 21st Iranian Conference on Electrical Engineering (ICEE-2013), Mashad, Iran, 2013.

11. N. Hajimohammadi and H. Bevrani, On load shedding design in microgrids. In: 18th Electric Power Distribution Conference, Kermanshah, Iran, April 30–May 1, 2013.

12. S. Shokoohi, H. Bevrani, and A. H. Naghshbandi, Application of neuro-fuzzy controller on voltage and frequency stability in islanded microgrids. In: Conference on Smart Electric Grids Technology (SEGT2012), Tehran, Iran, December 18–19, 2012, pp. 62–67.

13. A. H. Naghshbandi, S. Shokoohi, and H. Bevrani, Application of neuro-fuzzy controller on voltage and frequency stability in islanded microgrids, *J. Electr. Eng.*, **41**(2), 41–49, 2011.

14. M. Fathi and H. Bevrani, Adaptive price-based power flow in next generation electric power systems. In: International Symposium on Smart Grid Operation and Control (ISSGOC 2012), Sanandaj, Iran, 2012.

15. H. Bevrani and T. Ise, Virtual synchronous generators: a survey and new perspectives, Technical Report, Osaka University, Osaka, Japan, 2012.

16. P. Babahajyani, F. Habibi, and H. Bevrani, An on-line PSO-based fuzzy logic tuning approach: microgrid frequency control case study. In: International Symposium on Smart Grid Operation and Control (ISSGOC 2012), 2012.

17. F. Habibi, S. Shokoohi, and H. Bevrani, Designing a self-tuning frequency controller using ANN for an isolated microgrid. In: International Symposium on Smart Grid Operation and Control (ISSGOC 2012), 2012.

18. S. Shokoohi and H. Bevrani, PSO based droop control of inverter interfaced distributed generations. In: Conference on Smart Electric Grids Technology (SEGT2012), Tehran, Iran, December 18–19, 2012, pp. 77–82.

19. B. R. Barmish, *New Tools for Robustness of Linear Systems*, Macmillian, New York, 1994.

20. F. Karimi, E. J. Davison, and R. Iravani, Multivariable servomechanism controller forautonomous operation of a distributed generation unit: design and performance evaluation, *IEEE Trans. Power Syst.*, **25**(2), 853–865, 2010.

21. D. Xue, Y. Q. Chen, and D. P. Atherton, *Linear Feedback Control: Analysis and Design with MATLAB*, Vol. 14, Society for Industrial and Applied Mathematics, 2007.

22. H. Bevrani and T. Hiyama, *Intelligent Automatic Generation Control*, CRC Press, New York, 2011.

23. F. Katiraei and M. R. Iravani, Power management strategies for a microgrid with multiple distributed generation units, *IEEE Trans. Power Syst.*, **21**(4), 1821–1831, 2006.

APPENDIX A

NEW YORK/NEW ENGLAND 16-MACHINE 68-BUS SYSTEM CASE STUDY

Simplified two-axis generator model:

$$\dot{E}'_{qi} = \frac{1}{\tau'_{d0i}} \left[E_{FDi} - E'_{qi} + \left(x_{di} - x'_{di} \right) I_{di} \right] \tag{A.1}$$

$$\dot{E}'_{di} = \frac{1}{\tau'_{q0i}} \left[-E'_{di} + \left(x_{qi} - x'_{qi} \right) I_{qi} \right] \tag{A.2}$$

$$\dot{\omega}_i = \frac{1}{2H_i} \left[P_{mi} - \left(I_{di} E'_{di} + I_{qi} E'_{qi} \right) - D_i(\omega_i - \omega_s) \right] \tag{A.3}$$

$$\dot{\delta}_i = \omega_i - \omega_s \tag{A.4}$$

Power System Monitoring and Control, First Edition. Hassan Bevrani, Masayuki Watanabe, and Yasunori Mitani.
© 2014 John Wiley & Sons, Inc. Published 2014 by John Wiley & Sons, Inc.

TABLE A.1. Transmission Lines

Bus i	Bus j	R, pu	X, pu	B, pu
01	02	0.0035	0.0411	0.6978
01	30	0.0008	0.0074	0.4800
02	03	0.0013	0.0151	0.2572
02	25	0.0070	0.0086	0.1460
02	53	0.0000	0.0181	0.0000
03	04	0.0013	0.0213	0.2214
03	18	0.0011	0.0133	0.2138
04	05	0.0008	0.0128	0.1342
04	14	0.0008	0.0129	0.1382
05	06	0.0002	0.0026	0.0434
05	08	0.0008	0.0112	0.1476
06	07	0.0006	0.0092	0.1130
06	11	0.0007	0.0082	0.1389
06	54	0.0000	0.0250	0.0000
07	08	0.0004	0.0046	0.0780
08	09	0.0023	0.0363	0.3804
09	30	0.0019	0.0183	0.2900
10	11	0.0004	0.0043	0.0729
10	13	0.0004	0.0043	0.0729
10	55	0.0000	0.0200	0.0000
12	11	0.0016	0.0435	0.0000
12	13	0.0016	0.0435	0.0000
13	14	0.0009	0.0101	0.1723
14	15	0.0018	0.0217	0.3660
15	16	0.0009	0.0094	0.1710
16	17	0.0007	0.0089	0.1342
16	19	0.0016	0.0195	0.3040
16	21	0.0008	0.0135	0.2548
16	24	0.0003	0.0059	0.0680
17	18	0.0007	0.0082	0.1319
17	27	0.0013	0.0173	0.3216
19	20	0.0007	0.0138	0.0000
19	56	0.0007	0.0142	0.0000
20	57	0.0009	0.0180	0.0000
21	22	0.0008	0.0140	0.2565
22	23	0.0006	0.0096	0.1846
22	58	0.0000	0.0143	0.0000
37	65	0.0000	0.0033	0.0000
42	67	0.0000	0.0015	0.0000
52	68	0.0000	0.0030	0.0000
01	27	0.0320	0.3200	0.4100
23	24	0.0022	0.0350	0.3610
23	59	0.0005	0.0272	0.0000
25	26	0.0032	0.0323	0.5310
25	60	0.0006	0.0232	0.0000

TABLE A.1. (*Continued*)

Bus i	Bus j	R, pu	X, pu	B, pu
26	27	0.0014	0.0147	0.2396
26	28	0.0043	0.0474	0.7802
26	29	0.0057	0.0625	1.0290
28	29	0.0014	0.0151	0.2490
29	61	0.0008	0.0156	0.0000
09	30	0.0019	0.0183	0.2900
09	36	0.0022	0.0196	0.3400
36	37	0.0005	0.0045	0.3200
34	36	0.0033	0.0111	1.4500
35	34	0.0001	0.0074	0.0000
33	34	0.0011	0.0157	0.2020
32	33	0.0008	0.0099	0.1680
30	31	0.0013	0.0187	0.3330
30	32	0.0024	0.0288	0.4880
01	31	0.0016	0.0163	0.2500
33	38	0.0036	0.0444	0.6930
38	46	0.0022	0.0284	0.4300
46	49	0.0018	0.0274	0.2700
01	47	0.0013	0.0188	1.3100
47	48	0.0025	0.0268	0.4000
48	40	0.0020	0.0220	1.2800
35	45	0.0007	0.0175	1.3900
37	43	0.0005	0.0276	0.0000
43	44	0.0001	0.0011	0.0000
44	45	0.0025	0.0730	0.0000
39	44	0.0000	0.0411	0.0000
39	45	0.0000	0.0839	0.0000
45	51	0.0004	0.0105	0.7200
50	52	0.0012	0.0288	2.0600
50	51	0.0009	0.0221	1.6200
41	66	0.0000	0.0015	0.000

TABLE A.2. Generator Parameters

Gen i	MVA Base	x_l	x_d	x'_d	x''_d	T'_{do}	T''_{do}	x_q	x'_q	x''_q	T'_{qo}	T''_{qo}	H
1	300	0.003	0.969	0.248	0.147	12.6	0.045	0.6	0.25	0	0.035	0	3.4
2	800	0.035	1.8	0.42529	0.30508	6.56	0.05	1.7207	0.3661	0.30508	1.5	0.035	4.9494
3	800	0.0304	1.8	0.38309	0.32465	5.7	0.05	1.7098	0.36072	0.32465	1.5	0.035	4.9623
4	800	0.0295	1.8	0.29954	0.24046	5.69	0.05	1.7725	0.27481	0.24046	1.5	0.035	4.1629
5	700	0.027	1.8	0.36	0.27273	5.4	0.05	1.6909	0.32727	0.27273	0.44	0.035	4.7667
6	900	0.0224	1.8	0.35433	0.28346	7.3	0.05	1.7079	0.3189	0.28346	0.4	0.035	4.9107
7	800	0.0322	1.8	0.29898	0.24407	5.66	0.05	1.7817	0.27458	0.24407	1.5	0.035	4.3267
8	800	0.028	1.8	0.35379	0.27931	6.7	0.05	1.7379	0.31034	0.27931	0.41	0.035	3.915
9	1000	0.0298	1.8	0.48718	0.38462	4.79	0.05	1.7521	0.42735	0.38462	1.96	0.035	4.0365
10	1200	0.0199	1.8	0.48675	0.42604	9.37	0.05	1.2249	0.47929	0.42604	1.5	0.035	2.9106
11	1600	0.0103	1.8	0.25312	0.16875	4.1	0.05	1.7297	0.21094	0.16875	1.5	0.035	2.0053
12	1900	0.022	1.8	0.55248	0.44554	7.4	0.05	1.6931	0.49901	0.44554	1.5	0.035	5.1791
13	12000	0.003	1.8	0.33446	0.24324	5.9	0.05	1.7392	0.30405	0.24324	1.5	0.035	4.0782
14	10000	0.0017	1.8	0.285	0.23	4.1	0.05	1.73	0.25	0.23	1.5	0.035	3
15	10000	0.0017	1.8	0.285	0.23	4.1	0.05	1.73	0.25	0.23	1.5	0.035	3
16	11000	0.0041	1.8	0.35899	0.27809	7.8	0.05	1.6888	0.30337	0.27809	1.5	0.035	4.45

TABLE A.3. Excitation System Parameters

Gen i	k	T_7	T_6	T_5	E_{max}	$E_{R\,min}$
1	100	0.01	0	0	5	−5
2	100	0.01	0	0	5	−5
3	100	0.01	0	0	5	−5
4	100	0.01	0	0	5	−5
5	100	0.01	0	0	5	−5
6	100	0.01	0	0	5	−5
7	100	0.01	0	0	5	−5
8	100	0.01	0	0	5	−5
9	100	0.01	0	0	5	−5
10	100	0.01	0	0	5	−5
11	100	0.01	0	0	5	−5
12	100	0.01	0	0	5	−5
13	100	0.01	0	0	5	−5
14	100	0.01	0	0	5	−5
15	100	0.01	0	0	5	−5
16	100	0.01	0	0	5	−5

TABLE A.4. PSS Parameters

Gen i	K_{PSS}	T_1	T_2	T_3	T_4	U_{max}	U_{min}
01	100	0.10	0.02	0.08	0.02	0.2	−0.05
02	100	0.08	0.02	0.08	0.02	0.2	−0.05
03	100	0.08	0.02	0.08	0.02	0.2	−0.05
04	100	0.08	0.02	0.08	0.02	0.2	−0.05
05	100	0.08	0.02	0.08	0.02	0.2	−0.05
06	100	0.10	0.02	0.10	0.02	0.2	−0.05
07	100	0.08	0.02	0.08	0.02	0.2	−0.05
08	100	0.08	0.02	0.08	0.02	0.2	−0.05
09	100	0.08	0.03	0.05	0.01	0.2	−0.05
10	100	0.10	0.02	0.10	0.02	0.2	−0.05
11	100	0.08	0.03	0.05	0.01	0.2	−0.05
12	110	0.10	0.02	0.10	0.02	0.2	−0.05
13	100	0.10	0.02	0.10	0.02	0.2	−0.05
14	100	0.10	0.02	0.10	0.02	0.2	−0.05
15	100	0.10	0.02	0.10	0.02	0.2	−0.05
16	100	0.10	0.02	0.10	0.02	0.2	−0.05

NINE-BUS POWER SYSTEM CASE STUDY

TABLE B.1. Branch Data

From Bus	To Bus	R, pu	X, pu	B, pu	Limit, MW
1	4	0	0.0576	0	200
4	5	0.017	0.092	0.158	200
5	6	0.039	0.17	0.358	100
3	6	0	0.0586	0	250
6	7	0.0119	0.1008	0.209	100
7	8	0.0085	0.072	0.149	200
8	2	0	0.0625	0	200
8	9	0.032	0.161	0.306	200
9	4	0.01	0.085	0.176	200

Power System Monitoring and Control, First Edition. Hassan Bevrani, Masayuki Watanabe, and Yasunori Mitani.
© 2014 John Wiley & Sons, Inc. Published 2014 by John Wiley & Sons, Inc.

TABLE B.2. Generator Data

Generator	G1	G2	G3
Nominal power (MVA)	128	247.5	192
Type	Hydro	Steam	Steam
Speed (rpm)	180	3600	3600
V_{L-L} (KV)	16.5	18.3	13.8
X_d	0.146	0.8958	1.3125
X_d'	0.0608	0.1198	0.1813
X_d''	0.205	0.155	0.22
X_q	0.0969	0.8645	1.2578
X_q'	0.0969	0.1969	0.25
X_q''	0.221	0.143	0.292
X_L	0.0336	0.0521	0.0742
T_d'	8.96	6	5.89
T_d''	0.02	0.02	0.02
T_{qo}'	0.00002	0.535	0.6
T_{q0}''	0.02	0.02	0.02
R_S	2.8544e-3	2.8544e-3	2.8544e-3
H (on 100 MW)	23.64	3.01	6.4

Reactance values are in pu on 100 MVA base, and all generators are equipped with a governor and PSS.

TABLE B.3. Load Data

Load	Load A	Load B	Load C
Bus No.	9	5	7
Active power (MW)	125	90	100
Reactive power (MVAR)	50	30	35
Number of blocks	6	4	5

TABLE B.4. Exciter Data

Generator	Low Pass Filter	Regulator		Exciter		Output Limits	
	Tr	Ka	Ta	Ke	Te	Efmin	Efmax
G1	20e−3	200	0.001	1	0	0	7
G2	20e−3	200	0.001	1	0	0	12.3
G3	20e−3	200	0.001	1	0	0	12.3

APPENDIX C

FOUR-ORDER DYNAMICAL POWER SYSTEM MODEL AND PARAMETERS OF THE FOUR-MACHINE INFINITE-BUS SYSTEM

C.1 FOUR-ORDER LINEARIZED DYNAMICAL MODEL

In order to design a robust power system controller, it is first necessary to find an appropriate linear mathematical description of the multimachine power system. In the viewpoint of "generator unit i," the nonlinear state space representation model for such a system with two-axis model has the form

$$\dot{x}_{gi} = f\left(x_{gi}, u_{gi}\right) \tag{C.1}$$

where the states

$$x_{gi}^{\mathrm{T}} = \begin{bmatrix} x_{1gi} & x_{2gi} & x_{3gi} & x_{4gi} \end{bmatrix} = \begin{bmatrix} \delta_i & \omega_i & E'_{qi} & E'_{di} \end{bmatrix} \tag{C.2}$$

are defined as deviation from the equilibrium values

$$x_{egi}^{\mathrm{T}} = \begin{bmatrix} \delta_{1i}^e & \omega_{2i}^e & E_{qi}^{\prime e} & E_{di}^{\prime e} \end{bmatrix} \tag{C.3}$$

Power System Monitoring and Control, First Edition. Hassan Bevrani, Masayuki Watanabe, and Yasunori Mitani.
© 2014 John Wiley & Sons, Inc. Published 2014 by John Wiley & Sons, Inc.

The nonlinear model of (C.1) can be presented as follows:

$$\dot{x}_{1gi} = x_{2gi}$$

$$\dot{x}_{2gi} = -(D_i/M_i)x_{2gi} - (1/M_i)\Delta P_{ei}(x)$$

$$\dot{x}_{3gi} = -\left(1/T'_{d0i}\right)x_{3gi} - \left(\Delta x_{di}(x)/T'_{d0i}\right)\Delta I_{di}(x) + u_{gi}$$

$$\dot{x}_{4gi} = -\left(1/T'_{q0i}\right)x_{4gi} - \left(\Delta x_{qi}(x)/T'_{q0i}\right)\Delta I_{qi}(x)$$

(C.4)

where

$$\Delta P_{ei}(x) = \left(E'_{di}I_{di} + E'_{qi}I_{qi}\right) - \left(E'^{e}_{di}I^{e}_{di} + E'^{e}_{qi}I^{e}_{qi}\right)$$

(C.5)

$$I_{di} = \sum_{k}[G_{ik}\cos\delta_{ik} + B_{ik}\sin\delta_{ik}]E'_{dk}$$
$$+ \sum_{k}[G_{ik}\sin\delta_{ik} - B_{ik}\cos\delta_{ik}]E'_{qk}$$

(C.6)

and

$$I_{qi} = \sum_{k}[B_{ik}\cos\delta_{ik} - G_{ik}\sin\delta_{ik}]E'_{dk}$$
$$+ \sum_{k}[G_{ik}\cos\delta_{ik} + B_{ik}\sin\delta_{ik}]E'_{qk}$$

(C.7)

A detailed description of all symbols and quantities can be found in Ref. [1]. Using the linearization technique and after some manipulation, the nonlinear state equations (C.1) can be expressed in the form of following linear state space model.

$$\dot{x}_{gi} = A_{gi}x_{gi} + B_{gi}u_{gi}$$

(C.8)

where

$$A_{gi} = \begin{bmatrix} 0 & 1 & 0 & 0 \\ a_{21} & -\dfrac{D_i}{M_i} & a_{23} & a_{24} \\ a_{31} & 0 & a_{33} & -\dfrac{G_{ii}\Delta x_{di}}{T'_{d0i}} \\ a_{41} & 0 & \dfrac{G_{ii}\Delta x_{qi}}{T'_{q0i}} & a_{44} \end{bmatrix}, \quad B_{gi} = \begin{bmatrix} 0 \\ 0 \\ \dfrac{1}{T'_{d0i}} \\ 0 \end{bmatrix}$$

(C.9)

and

$$\Delta x_{di} = x_{di} - x'_{di}, \quad \Delta x_{qi} = x_{qi} - x'_{di}$$

(C.10)

The above parameters are defined as follows:

δ_i Machine rotor angle
ω_i Machine rotor speed
E'_{di} d-axis internal machine voltage
E'_{qi} q-axis internal machine voltage
D_i Damping constant
M_i Inertia constant
G_{ii} Driving point conductance
T'_{doi} d-axis open-circuit transient time constant
T'_{qoi} q-axis open-circuit transient time constant
x_{di} d-axis synchronous reactance
x'_{di} d-axis transient reactance
x_{qi} q-axis synchronous reactance

The elements of A_{gi} matrix in (C.10) are

$$a_{21} = -\frac{1}{M_i}\frac{\partial f_{1i}(x)}{\partial x_{1gi}}\bigg|_{x_{egi}} \tag{C.11}$$

$$a_{23} = -\frac{\left[G_{ii}E'^e_{qi} - B_{ii}E'^e_{di} + I^e_{qi}\right]}{M_i} - \frac{1}{M_i}\frac{\partial f_{1i}(x)}{\partial x_{3gi}}\bigg|_{x_{egi}} \tag{C.12}$$

$$a_{24} = -\frac{\left[G_{ii}E'^e_{di} + B_{ii}E'^e_{qi} + I^e_{di}\right]}{M_i} - \frac{1}{M_i}\frac{\partial f_{1i}(x)}{\partial x_{4gi}}\bigg|_{x_{egi}} \tag{C.13}$$

$$a_{31} = -\frac{\Delta x_{di}}{T'_{d0i}}\frac{\partial f_{2i}(x)}{\partial x_{1gi}}\bigg|_{x_{egi}}, \quad a_{33} = -\frac{1}{T'_{d0i}} + \frac{B_{ii}\Delta x_{di}}{T'_{d0i}} \tag{C.14}$$

$$a_{41} = -\frac{\Delta x_{qi}}{T'_{q0i}}\frac{\partial f_{3i}(x)}{\partial x_{1gi}}\bigg|_{x_{egi}}, \quad a_{44} = -\frac{1}{T'_{q0i}} + \frac{B_{ii}\Delta x_{qi}}{T'_{q0i}} \tag{C.15}$$

where

$$f_{1i}(x) = x_{4gi}\Delta I_{di}(x) + x_{3gi}\Delta I_{qi}(x) + \sum_{k \neq i}\left\{\left[E'^e_{di}\eta_{ik}(\delta) + E'^e_{qi}\hat{\eta}_{ik}(\delta)\right]x_{4gk}\right.$$
$$+ \left[E'^e_{di}\nu_{ik}(\delta) + E'^e_{qi}\hat{\nu}_{ik}(\delta)\right]x_{3gk} + \left.\left[E'^e_{di}\nu_{ik}(\delta) + E'^e_{qi}\hat{\nu}_{ik}(\delta)\right]\sin\phi_{ik}\right\} \tag{C.16}$$

$$f_{2i}(x) = \sum_{k \neq i}\left[\eta_{ik}(\delta)x_{4gk} + \nu_{ik}(\delta)x_{3gk} + \nu_{ik}(\delta)\sin\phi_{ik}\right] \tag{C.17}$$

$$f_{3i}(x) = \sum_{k \neq i} \left[\hat{\eta}_{ik}(\delta) x_{4gk} + \hat{\nu}_{ik}(\delta) x_{3gk} + \hat{\upsilon}_{ik}(\delta)\sin \phi_{ik} \right] \tag{C.18}$$

$$\eta_{ik}(\delta) = G_{ik} \cos \delta_{ik} + B_{ik} \sin \delta_{ik}, \ \hat{\eta}_{ik}(\delta) = B_{ik} \cos \delta_{ik} - G_{ik} \sin \delta_{ik} \tag{C.19}$$

$$\nu_{ik}(\delta) = G_{ik} \sin \delta_{ik} - B_{ik} \cos \delta_{ik}, \ \hat{\nu}_{ik}(\delta) = B_{ik} \sin \delta_{ik} - G_{ik} \cos \delta_{ik} \tag{C.20}$$

$$\upsilon_{ik}(\delta) = 2g1_{ik} \sin \frac{\delta_{ik}^e + \delta_{ik}}{2} + 2g2_{ik} \cos \frac{\delta_{ik}^e + \delta_{ik}}{2}, \ \phi_{ik} = 0.5\left(x_{1gi} - x_{1gk}\right) \tag{C.21}$$

$$\hat{\upsilon}_{ik}(\delta) = 2g2_{ik} \sin \frac{\delta_{ik}^e + \delta_{ik}}{2} - 2g1_{ik} \cos \frac{\delta_{ik}^e + \delta_{ik}}{2}, \ \delta_{ik} = \delta_i - \delta_k \tag{C.22}$$

$$g1_{ik} = G_{ik}E_{dk}'^e - B_{ik}E_{qk}'^e, \ g2_{ik} = G_{ik}E_{qk}'^e + B_{ik}E_{qk}'^e \tag{C.23}$$

C.2 FOUR-MACHINE INFINITE-BUS POWER SYSTEM

TABLE C.1. Generator Constants

Unit No.	M_i, s	D_i	x_{di}, pu	x_{di}', pu	x_{qi}, pu	x_{qi}', pu	T_{d0i}', s	T_{q0i}', s	MVA
1	8.05	0.002	1.860	0.440	1.350	1.340	0.733	0.0873	1000
2	7.00	0.002	1.490	0.252	0.822	0.821	1.500	0.1270	600
3	6.00	0.002	1.485	0.509	1.420	1.410	1.550	0.2675	1000
4	8.05	0.002	1.860	0.440	1.350	1.340	0.733	0.0873	900

TABLE C.2. Line Parameters

Line No.	Bus–Bus	R_{ij}, pu	X_{ij}, pu	S_{ij}, pu
1	1–9	0.02700	0.1304	0.0000
2	2–10	0.07000	0.1701	0.0000
3	3–11	0.04400	0.1718	0.0000
4	4–12	0.02700	0.1288	0.0000
5	10–6	0.02700	0.2238	0.0000
6	11–7	0.04000	0.1718	0.0000
7	12–8	0.06130	0.2535	0.0000
8	9–10	0.01101	0.0829	0.0246
9	10–11	0.01101	0.0829	0.0246
10	11–12	0.01468	0.1105	0.0328
11	12–5	0.12480	0.9085	0.1640

TABLE C.3. Excitation Parameters

K_1	K_2	K_3	K_4	$\left|E_{max(min)}\right|$	U_{max}	U_{min}
1.00	19.21	10.00	6.48	5.71	7.60	−5.20
T_1 (s)	T_2 (s)	T_3 (s)	T_4 (s)	T_5 (s)	T_6 (s)	T_7 (s)
0.010	1.560	0.013	0.013	0.200	3.000	10.000

TABLE C.4. Governor and Turbine Parameters

Parameters	Unit 1	Unit 2	Unit 3	Unit 4
T_1 (s)	0.08	0.06	0.07	0.07
T_2 (s)	0.10	0.10	0.10	0.10
T_3 (s)	0.10	0.10	0.10	0.10
T_4 (s)	0.40	0.36	0.42	0.42
T_5 (s)	10.0	10.0	10.0	10.0
T_H (s)	0.05	0.05	0.05	0.05
T_I (s)	0.08	0.08	0.08	0.08
T_L (s)	0.58	0.58	0.58	0.58
K_H (pu)	0.31	0.31	0.31	0.31
K_I (pu)	0.24	0.24	0.24	0.24
K_L (pu)	0.45	0.45	0.45	0.45
M_1 (pu/min)	0.50	0.50	0.50	0.50
M_2 (pu/min)	0.20	0.20	0.20	0.20
M_3 (pu/min)	1.50	1.50	1.50	1.50
N_1 (pu/min)	−0.50	−0.50	−0.50	−0.50
N_2 (pu/min)	−0.20	−0.20	−0.20	−0.20
N_3 (pu/min)	−0.50	−0.50	−0.50	−0.50

TABLE C.5. Conventional PSS Parameters

T_r, s	G_{PSS}	U_{max}, pu	T_1, s
5.00	10.00	1.00	0.025
T_2 (s)	T_3 (s)	T_4 (s)	T_5 (s)
0.056	0.054	0.037	0.53

REFERENCE

1. P. W. Sauer and M. A. Pai, *Power System Dynamic and Stability*, Prentice-Hall, Englewood Cliffs, NJ, 1998.

INDEX

ACE. *See* Area control error (ACE)

AC MG frequency response model, 222

Adaptive neuro-fuzzy inference system (ANFIS), 216–221

  closed-loop control system for VSI-based inverter, 219

  generalized droop control concept, 218, 219

  inputs and outputs, 218

  structure of, 217

  trained network output *vs.* real output, 219

  VSI inverter-based MG, 220

AGC. *See* Automatic generation control (AGC)

Algebraic Riccati equation (ARE), 124

Algorithms, 47, 52, 56, 57, 138, 140, 148

  artificial neural network (ANN), 217

  back-propagation (BP), 217

  ECS (ACA-ECS) algorithm, 237

  genetic, 89, 97, 101

  iterative linear matrix inequalities (ILMI), 126, 128–130

  LS algorithm, 52, 57, 64, 81, 82, 159–161, 203, 243–246

  PSO, 224, 226

  statistical cooperative power dispatching (SCPD), 241, 242

  UFLS/UVLS, 160, 166, 176

Analog network simulator (ANS), 126

ANFIS. *See* Adaptive neuro-fuzzy inference system (ANFIS)

Angle and voltage control, 73, 74

Angle–voltage deviation graph, 45–48

Angular velocity, 8, 28, 29, 98, 99

ANN. *See* Artificial neural network (ANN)

ANN-based self-tuning frequency control, 228

  in intelligent control, 229–230

  proposed control scheme, 230–234

    ANN structure used for PI tuning, 231

    system response, 234

    updating weights via back-propagation learning, 233

Area control error (ACE), 78–80, 82, 94

  calculation, 78

  effects of local load changes and interface, 78

  signals, filtered before, 79

Artificial neural network (ANN), 217, 228–231, 234

  intelligent control structures, 230

  MG test system, 232

  PI tuning, 231

Artificial neural networks, 89

Automatic frequency and voltage controls, 72

Automatic generation control (AGC), 75, 82, 84, 86, 94

  communicate with SCADA, 84

  frequency-dependent reserves, activated by, 82, 83

Automatic voltage regulators (AVRs), 32, 73, 119

  challenge/limitations, 126

  conventional, 120

  direct feedback linearization (DFL) is used to design, 124

  integrated design approach, 123

  intelligent approaches, 125

Power System Monitoring and Control, First Edition. Hassan Bevrani, Masayuki Watanabe, and Yasunori Mitani.

Automatic voltage regulators (AVRs)
 (*Continued*)
 internal model control (IMC), 123
 to construct a robust controller, 123
 model predictive control (MPC) method,
 124
 particle swarm optimization (PSO)
 technique, 125
 resources, 123
 switching-based coordination method, 124
 switching concept-based coordination
 method, 124
 trapezoid-shaped membership functions,
 125
 weighed switching-based coordination
 method, 125
AVR and PSS coordination design, 120–123,
 149
 fuzzy logic-based coordination system,
 149–151
 membership functions for signals, 151
 simulation results, 151–155
 system generator response, 153–154
AVR–PSS synthesis approach, robust
 simultaneous, 126
 control framework, 126–128
 control structure, 127
 proposed $H\infty$-SOF control framework,
 127–128
 developed algorithm, 128–130
 BMI problems, 128
 ILMI algorithm, 129–130
 stabilizing SOF gain matrix, 128
 experiment results, 132–137
 real-time implementation, 131–132
 control/monitoring desks, 133
 excitation control system, 132
 four-machine infinite bus power system,
 131
 performed computer-based control loop,
 133
 speed governing and turbine system, 132
AVRs. *See* Automatic voltage regulators
 (AVRs)

Back-propagation learning, 230, 232, 233
Battery energy storage system (BESS), 221,
 225, 233
Bias factor, 77, 78

Case studies
 isolated MG, 210, 221
 new York/New England 16-machine 68-bus
 system, 249–253
 nine-bus power system, 254–255
 and simulation, 243–246
 SSR phenomenon, and increasing AVR
 gains, 148
Circuit breaker (CB), 210
Communications
 and application levels, 4
 between control units, 87
 data, 86
 distribution systems, 192
 EMS-leve, 87
 infrastructures, 206
 within the grid, 240
 internet protocol (IP) based, 206
 low-bandwidth, 198, 202
 narrowband, 5
 power distribution networks with two-way,
 235
 power line communication (PLC), 206
 SCADA-level, 87
 servers, 86
 speed and accuracy, 3
 standard protocols, 85
 wideband, 4, 5
Contingency reserve services, 83
Controller tuning, using a vibration model, 98
 simulation results, 102–107
 comparison of eigenvalues variation,
 105, 107
 exciter model with $\Delta\omega$ type PSS, 103
 result of PSSs tuning in the nighttime
 condition, 107
 result of the PSS tuning in daytime
 condition, 105
 simulation with tuned PSSs of
 generators, 104, 108
 stability change with gradual tuning of
 PSS connected to generator, 106
 tuning mechanism, 101–102
 vibration model, including effect of
 damping controllers, 98–100
Control synthesis. *See also* Microgrids (MGs)
 global control synthesis
 adaptive energy consumption scheduling,
 235–236

connected MGs, distribution network, 236–237
proposed methodology, 237–240
power dispatching in interconnected MGs, 240
methodology, 241–242
local control synthesis, 209
droop-based voltage (*See* Intelligent droop-based voltage)
frequency control, 215–216
robust voltage control design, 209
case study, 210–213
controller design, 213–215
secondary control synthesis, 221
conventional/fuzzy PI-based frequency control, 222–224
fuzzy PI-based secondary frequency control, 223
fuzzy rules set, 224
PI control parameters, 223
intelligent frequency control, 221–222
AC MG frequency response model, 222
case study, 221–222
PSO-fuzzy PI frequency control, 224
closed-loop system, 225
online PSO algorithm, 226
simulation results, 224–228
Coordinated AVR-PSS, 135, 141

Damping coefficient, 8, 28, 29, 77, 79, 94, 225
Damping controllers, 96–98, 107, 109
Data acquisition, 2, 26, 81, 83–88
Data collection, 5
Data exchange, 5, 31, 86
Defuzzification layer, 149, 151, 217
DFIGs. *See* Doubly fed induction generators (DFIGs)
Diesel engine generator (DEG), 221, 222, 225
frequency deviation, 229, 233
power response, 234
Diesel generators, 188, 200, 201
Distributed generator (DG), 65, 70, 90, 210, 242
Disturbance/fault analysis, 84
Doubly fed induction generators (DFIGs), 90–92, 91, 136, 161–164, 170, 171, 175, 195
Dynamic controller, 77

Eastern interconnected system, 3
ECSs. *See* Energy capacitor systems (ECSs)

Electromechanical dynamics, 8
Elkraft Power Co., 3
Emergency control synthesis, 242
developed LS algorithm, 244
case studies, 243–246
ROCOF, 243
simulation, 243–246
system response after using LS scheme, 245
Emergency protection/control loop, 80
EMS. *See* Energy management system (EMS)
Energy capacitor systems (ECSs), 188, 200–202, 235–237, 239
Energy consumption scheduling, 235
Energy imbalance management, 82
reserve, 83
Energy management system (EMS), 84–87, 203
application layer, 84
EUROSTAG software, 103, 111
Excitation system, 49, 129, 131, 253
control system, 120

FACTS. *See* Flow controlling reference points (FACTS)
Fixed speed induction generators (FSIGs), 138
Florida Reliability Coordinating Council (FRCC) standard, 164
Flow controlling reference points (FACTS), 27, 72, 73
Flywheel energy storage system (FESS), 221, 225, 233
Four-machine infinite-bus power system, 259–260
conventional PSS parameters, 260
excitation parameters, 259
generator constants, 259
governor/turbine parameters, 260
line parameters, 259
Four-order linearized dynamical model, 256–259
Frequency
control, 75–77, 215, 216
dynamic, 77–81
FC and DEG, 229
load disturbance, 227
operating states, 81–83
power reserves, 81–83
controlled protection devices, 80
dependent reserves, 82
deviations, 79, 81

Frequency (*Continued*)
  ranges and associated control actions, 82
  and rate of frequency change
    G2 outage with/without WTGs, 164
  generation–load dynamic, relationship, 79
  monitoring network, 3
  regulation, with contribution of RESs, 93
  response model, 93
  stability, 71, 72
  variation, 81
Fuel cell (FC) system, 195, 200, 221
  frequency deviation, 229
  fuel blocks, 222
Fuzzy PI-based secondary frequency control, 223
Fuzzy rules, 223–224

Global positioning system (GPS), 2
  satellite clock, 87
  synchronized GPS time, 4
Graphical tools
  angle–voltage deviation graph, 45–48
  electromechanical wave propagation graph, 60–62
    angle wave and system configuration, 64–67
    wave propagation, 62–64
  frequency–angle deviation graph, 58–60
  simulation results, 48–49
    disturbance in demand side, 50, 52
    disturbance in generation side, 49–50
  voltage–frequency deviation graph, 52–53
  Δv–Δf graph
    for contingency assessment, 53–56
    for load shedding synthesis, 56–57
  in WAMS, importance of, 43–45

High-voltage DC (HVDC) special controls, 71
Hurwitz criteria, polynomials, 213, 214
Hysteresis modulation, 188

ICT architecture, 4
  data collection, 5
    utility-owned communications, 5
  data exchange between utilities require, 5
  requirements for a PDC, 4
  schema for wide-area phasor measurement system, 4
    PMU/WAMS, 4

standards of phasor data transmission
    protocol, 4
  IEEE C37.118, 4
storage of data archiving for, 5
  data files, 5
IEEE nine-bus power system, 162
  updated version, 162
Independent power producers (IPPs), 12
Induction generators (IGs), 90, 161, 195
Inertia, 20, 28, 29, 30, 49, 76, 77, 139, 144, 161, 225
Information and communication technology (ICT), 1
Instantaneous contingency reserves, 83
Intelligent droop-based voltage, 215–221
  neuro-fuzzy control framework, 217–218
    structure of ANFIS, 217
  synthesis procedure/performance evaluation, 218–221
    inputs/outputs of proposed controller, 218–219
    load change pattern scenario, 220
    trained network output *versus* real output, 219
    VSI inverter-based MG with, 220
Intelligent systems in real power systems, 88, 89
  problems for future extension, 89
Intelligent technologies, in Japanese power system, 89
Interarea low-frequency oscillations, 7, 14, 15, 20, 27, 30, 31, 99, 112, 116, 117
International Council on Large Electric Systems (CIGRE), 2, 84
  technical report dealing with, 2
Interutility control center communication protocol (ICCP), 85
Inverter terminal voltage, 211

Kharitonov's theorem, 209, 213, 216
Kinetic energy, 81, 181
Knowledge-based expert systems, 89

Laplace transform, 79, 81
Lead-lag compensator, 74
Least-squares error (LSE) method, 217
Load change pattern scenario, 220
Load controllers (LC), 188
Load demand management, 240

Load disturbances, 52, 54, 55, 79, 80, 94, 212, 215, 224, 225, 232, 234, 243
  multiple step, 227
Load-frequency control (LFC), 75
  model, 79
Load–generation system, 82
Load–generation variation, 83
Load shedding (LS), 52, 57, 64, 81, 82, 158–161, 159–161, 203, 243, 246
  P–Q coupling, 164, 165
  simultaneous voltage and frequency-based, 165
    approach for optimal UFVLS, 176–177
    case studies, 168–170
      24-bus test system, 170–176
      nine-bus test system, 170
    implementation, 167–168
    proposed LS scheme, 165–167
    robustness of UFVLS in presence of WTGs, 177–178
    simulation results, 168–170, 177
  standards, 164
  system response with, 165
Local area network (LAN), 85
Local control synthesis
  robust voltage control design
  controller design
    acceptable values, of polynomials, 215
    Kharitonov's theorem, application, 213–214
Long-term stability problem, 72
Low-frequency oscillation dynamics, 16
  electromechanical modes characteristics, 16–18
    low-frequency oscillations, mode1/2, 20
    phase differences, 18
    Fourier spectrum of, 19
  Southeast Asia power network
    oscillation characteristics analyses, 18–23
    frequency analysis, 22
    frequency deviations during, 21
    low-frequency oscillations, about 0.5 Hz/0.3 Hz, 23
LS. *See* Load shedding (LS)

Mamdani fuzzy inference system, 223
Mathematical models, 89
  limitations, 89

MATLAB software, 218
Maximum power point tracking (MPPT), 195
Measurement-based controller design, 97
MG central controller (MGCC), 188, 193, 198, 202, 203, 205
Microgrids (MGs), 76, 186–192
  architecture, 187
  control, 192–194
    central/emergency controls, 204–206
    general scheme, 193
    global controls, 202–204
    local, 195–197
    loops, 194
    secondary, 198–202
  model for microsource connected to, 188
Multiarea power system, 77

National Electrical Reliability Coordinating Council (NERCC), 160
National Electricity Market Management Company, 3
Neuro-fuzzy control framework, 217–218
  design steps, 217, 218
  structure of ANFIS, 217
Nine-bus power system, 254–255
Nine-bus test system, 91
Nonspinning (contingency) reserve, 82, 83

Off-line stability evaluation, 84
Oscillation characteristics, in power systems, 8
  Eigenvalue analysis, 8–9
  in interconnected power system, 9–11
    block diagram of AVR, 10
    generator rated capacity and output, 10
    IEEJ WEST 10-machine system model, 9
    mode shapes associated with generator angle, 11
    participation factors associated with generator angle, 11
    system constants of WEST 10-machine system, 10
  participation factor, 8–9
Oscillation model identification
  using phasor measurements, 29
    dominant mode identification, with signal filtering, 30–32
    oscillation model of electromechanical mode, 29–30

Oscillation monitoring, using phasor
measurements, 12
Japan Power Network, monitoring of, 12–14
NCT2000, 12
overall schema for the campus WAMS, 14
phasor measurement unit, 13
phasor voltage, computation using
sinusoidal voltage, 13–14
monitoring of the Southeast Asia power
network, 14–15
WAMS for power system dynamics
using PMUs, 15
Oscillations
in accelerator powers, 152
angular velocity of, 29
characteristics (*See* Oscillation
characteristics, in power systems)
damping, 215
coordinator designs, 125
with poor, 3
damp interarea, 99
dynamics, 7, 41, 107
low-frequency, 16–23, 38, 99, 106, 109,
112, 113, 116, 120
electromechanical, 120
interarea low-frequency, 7, 14, 15, 20, 27,
30, 31, 99, 112, 116, 117
magnitude of, 175, 177
model identification using (*See* Oscillation
model identification)
modes, 131, 134, 137
monitoring
interarea, 4
using phasor measurements
(*See* Oscillation monitoring)
original and filtered, 34
power, 98
second-order oscillation model, 108
small-signal, 101
stability, 3
torsional, 121
undesirable, 162
voltage/rotor angle, 144
Overfrequency generation trip (OFGT), 80
Overfrequency generator shedding (OFGS),
80

Participation factors, 80
associated with generator angle, 11

Particle swarm optimization (PSO) techniques,
221
PDCs. *See* Phasor data concentrators (PDCs)
Peak-to-average ratio (PAR), 235
Phase angle, 98
Phasor data concentrators (PDCs), 4, 5
Phasor measurement units (PMUs), 7
to collect the data measured by, 4
data automatically collected by PDCs, 13
frequency variation observed by, 58
installations, 2, 14
installed within service area of each power
company, 13
measurements, 58
PMU/intelligent electronic devices (IEDs),
85
projects in Canada, 3
protocol applies to sending data from, 5
schema of PMU/WAMS, 4
Photovoltaic (PV) panel, 221
PI feedback control, 214
PMUs. *See* Phasor measurement units (PMUs)
Point of common coupling (PCC) junction,
210, 211
Power dynamic management, 3
Power generation–load balance, 80
Power grids, 2, 76, 90, 161, 168, 170, 210, 241
PowerLink Co., 3
Power Log, 3
Power system
controls
emergency controls, 72, 76
normal/preventive controls, 72
interconnected, 79
oscillation dynamics, 7
parameters, 225
stability, 70 (*See also* Power system
stabilizers (PSSs))
defined, 71
small-signal, 27–29
Power system stabilizers (PSSs), 96, 99, 119
challenge/limitations, 126
conventional, 120
direct feedback linearization (DFL) is used
to design, 124
integrated design approach, 123
intelligent approaches, 125
internal model control (IMC), 123
to construct a robust controller, 123

model predictive control (MPC) method, 124
particle swarm optimization (PSO) technique, 125
resources, 123
switching-based coordination method, 124
switching concept-based coordination method, 124
trapezoid-shaped membership functions, 125
weighed switching-based coordination method, 125
Primary control, 75, 76, 80. *See also* Frequency, control
PSS–AVR system, 73
PSSs. *See* Power system stabilizers (PSSs)
Pulse density modulation (PDM), 188
Pulse width modulation (PWM), 188

Rate of change of frequency (ROCOF), 243
Real-time monitoring, 8
Real-time stability assessment, 84
Real-world power systems, 77
Regulation power, 201
reserve, 82, 93
Reliability test system (RTS), 170, 182
Remote terminal units (RTUs), 85
Renewable energy source (RES), 1, 89, 135
actual RES tie-line power, 94
contribution to regulation services, 92–94
frequency regulation with contribution of, 93
impacts of renewable energy options, 90–92
power fluctuation, 94
variations caused by, 90
Replacement reserves, 83
RES. *See* Renewable energy source (RES)
Research Laboratory of the Kyushu Electric Power Company (KEPCO), 126
Rotating inertia, 191
Rotor angle stability, 71, 72

SCADA system. *See* Supervisory control and data acquisition (SCADA) system
Secondary control, 75, 78, 81, 82, 200, 202
Secondary voltage control, 74
Security assessment, and control system, 84
Short-term stability problem, 72
Simulation, 91
case study, 243–246
data and system parameters, 91, 162

model, 9
package, 97, 103
results, 48, 57, 65, 92, 102, 141, 148, 151, 163, 165, 168, 171–173, 176, 177, 180, 215, 224, 238, 241
sampling time, 218
time domain, 103
Small-signal stability assessment, 32
based on frequency monitoring, 38–41
comparison of eigenvalues, 40–41
comparison of modeling accuracy, 39
FFT results, 38
filtered speed deviations of each point, 38
based on phasor measurements, 33–37
comparison of eigenvalues, 37
comparison of modeling accuracy, 36
FFT results, 35
location of installed PMUs, 35
simulation study, 32–33
comparison of Eigenvalues with type of filtering method, 35
eigenvalues of dominant modes, calculated by, 32
generator rated capacity, 33
original and filtered oscillations, 34
Smart grid technologies, 2
Spare power capacity, 82
Spinning reserve, 82, 83
Standards redesign, 90
Static compensator (STATCOM), 91, 162–164
Static VAR compensator (SVC), 48, 72, 138, 140, 141, 144, 148
Statistical cooperative power dispatching (SCPD), 241, 242
Substation automation systems (SASs), 85
Supervisory control and data acquisition (SCADA) system, 2, 84–88, 85, 86, 94, 179
architecture, 88
regional SCADA, West Regional Electric Co., 87
structure, 85
typical center, simplified structure, 86
Sustainable energy, 1
SVC. *See* Static VAR compensator (SVC)
Synchronized phasor measurement, 2
Synchronous condensers, 72, 171
Synchrophasors, IEEE Standard, 2

System parameters
   open-loop bode diagram, 213, 214
   rated values, 212
System response. *See various algorithm*
   applications

Tabu search, 89
Tertiary voltage control, 74, 76
Torsional oscillation, 121
Trade-off
   AVR and PSS performance, 124
   optimal (full NAs demand) and uniform
      (no NAs demand) schemes, 240
   simplicity and preciseness, 178
   voltage regulation and closed-loop stability,
      127
Tuning mechanism, 101–102
Turbine time constant, 225

UFVLS scheme. *See* Underfrequency/voltage
   LS (UFVLS)
Uncertain parameters
   variation range, 227, 228
Underfrequency generation trip (UFGT), 80
Underfrequency load shedding (UFLS), 52,
      56, 57, 72, 77, 80, 81, 158, 159,
      160, 164, 166, 170, 173, 174, 177,
      183, 205
   strategy, 80
Underfrequency/voltage LS (UFVLS), 166, 168
   considering time delay and permitted
      shedding region, 57, 167
   flowchart, 169
   optimal approach, 176–177
   robustness of UFVLS in the presence of
      WTGs, 177–178
Undervoltage load shedding (UVLS), 52, 56,
      72, 158, 159, 160, 165, 173, 177,
      205
Utility-owned communications, 5

Vibration model, 98
   dynamics, 98–99
Virginia Tech, 3
Voltage behavior
   impact of wind turbines on, 162
      DFIG-type penetration, 162–163
      IG-type penetration, 162–163
      with/without static compensator,
         162–163

Voltage compensators, 74
Voltage control design, 209–215
   case study, 210–213
   designs, 213–215
Voltage stability, 71, 73, 74, 161
Voltage variation, 47, 63, 74
VSI-based inverter, 219

WAMS. *See* Wide-area measurement system
      (WAMS)
Wave propagation-based emergency control,
      178
   proposed control scheme, 178–180
      framework of proposed methodology,
         179
   simulation results, 180
      application to 24-bus test system,
         182–183
      ring systems and islanding, 180–182
Western Electricity Coordinating Council,
      United States, 3
Wide-area measurement system (WAMS),
      1–2
   based controller design, 107
      design procedure, 110
         closed-loop system configuration, 110
      simulation results, 110–117
         Bode diagram of the model, 114
         coefficients of $H\infty$ controller, 115
         comparison of bode diagrams, 116
         eigenvalues of simple two-area system,
            116
         exciter and PSS model for generators,
            111
         filtered phase difference, 113
         filtered phase difference and output
            signal, 115
         Fourier spectrum, 112
         input signal u for low-order model, 114
         with original two-area system, 117
         phase difference between nodes, 112
         simple two-area system model, 111
         simulation with tuned PSSs of
            generators, 108
         step response with low-order model,
            117
      wide-area power system identification,
         107–109
         approximate oscillation model of
            interested mode represented by, 109

Wide-area measurement system (WAMS)
  (*Continued*)
    oscillation dynamics with a single
      mode, 107
    procedure of determining model *F(s)*,
      109
    simulation with tuned PSSs of
      generators, 108
  based coordination approach, 135
    application, 141
    developed algorithm, 138–141
      control laws, 139
      generalized stable region of power
        system, 138
      proposed control algorithm, 140
      rotor rotation, 139
      swing equation, 139
    high penetration of wind power, 136–138
    simulation results, 141–149
      with wind power, 144, 146–147
      without wind power, 141–145
  based interarea mode identification, 15–16
    fast Fourier transform (FFT)-based filter,
      for pass band, 16
    original and filtered phase differences, 17
  damping controller design with, 97

emergency control (*See* Load shedding (LS))
  performed Campus WAMS
    configuration in Japan, 12
    overall framework, 14
  small-signal stability assessment (*See*
      Small-signal stability assessment)
  tuning mechanism, 101, 102
Wide-area monitoring system. *See* Wide-area
      measurement system (WAMS)
Wide-area phasor measurement system, 4,
      8, 15, 96, 97, 101, 106, 107, 117
Wide-area power system dynamics, 3
Widrow–Hoff correlation, 230
Wind
  energy, 200
  generation, 3
  power
    facilities, 93
    high penetration, 48, 136–138, 161, 167,
      171
    on system performance, 90
  speed, 170, 177
  turbine generator, 57, 90, 158, 161, 163,
      170, 175, 177, 221, 222, 232–234

Ziegler–Nichols method, 222, 223